THE ANIMAL WORLD

The Indian leopard *(Panthera pardus fusca)* is one of the most endangered of all the big cats.

THE WORLD BOOK ENCYCLOPEDIA OF SCIENCE
VOLUME 6

THE ANIMAL WORLD

WORLD BOOK, INC.
a Scott Fetzer company

CHICAGO

Staff

President
Robert C. Martin

**Vice President and
Editor in Chief**
Michael Ross

Editorial

Managing Editor
Maureen Mostyn Liebenson

Writers
Bonny Davidson
Karen Ingebretsen
Rita Vander Meulen

Permissions Editor
Janet T. Peterson

Indexer
David Pofelski

**Executive Director of
Research and Product
Development**
Paul A. Kobasa

Researchers
Lynn Durbin
Cheryl Graham
Karen McCormack
Loranne Shields

Consultant
Steve Thompson
Director of
Conservation and Science
Lincoln Park Zoo

Art

Executive Director
Roberta Dimmer

Art Director
Wilma Stevens

Senior Designer
Isaiah Sheppard

Cover Design
Chestnut House

Photography Manager
Sandra Dyrlund

Product Production

**Senior Manager, Pre-press
and Manufacturing**
Carma Fazio

Manager, Manufacturing
Barbara Podczerwinski

Senior Production Manager
Madelyn Underwood

**Manufacturing Production
Assistant**
Valerie Piarowski

Proofreaders
Anne Dillon
Chad Rubel

Text Processing
Curley Hunter
Gwendolyn Johnson

World Book, Inc.
233 N. Michigan Ave.
Chicago, IL 60601

For information on sales to schools and libraries, call 1-800-WORLDBK (967-5325), or visit our Web site at http://www.worldbook.com

Library of Congress Catalog Card No. 00-109505
ISBN: 0-7166-3399-X (set)
ISBN: 0-7166-3355-8 (vol. 6)
Printed in the United States of America

4 5 6 7 8 9 06 05 04 03 02 01 00

Contents

Preface

The quest to explore the known world and to describe its creation and subsequent development is nearly as old as humankind. In the Western world, the best known creation story comes from the book of Genesis. It tells how God created the earth and all living things. Modern religious thinkers interpret the Biblical story of creation in various ways. Some believe that creation occurred exactly as Genesis describes it. Others think that God's method of creation is revealed through scientific investigation. **The Animal World** presents an exciting picture of what scientists have learned about the great variety of the Earth's animal life.

The editorial approach

The object of the *Encyclopedia of Science* is to explain for adults and children alike the many aspects of science that are not only fascinating in themselves but are also vitally important for an understanding of the world today. To achieve this, the books in this series are straightforward and concise, accurate in content, and are clearly and attractively presented.

The often forbidding appearance of traditional science publications has been completely avoided in the *Encyclopedia of Science*. Approximately equal proportions of illustrations and text make even the most unfamiliar subjects interesting and attractive. Even more important, all of the drawings have been created specially to complement the text, each explaining a topic that can be difficult to understand through the printed word alone.

The thorough application of these principles has created a publication that covers its subject in an interesting and stimulating way, and that will prove to be an invaluable work of reference and education for many years to come.

The advance of science

One of the most exciting and challenging aspects of science is that its frontiers are constantly being revised and extended, and new developments are occurring all the time. Its advance depends largely on observation, experimentation, and debate, which generate theories that have to be tested and even then stand only until they are replaced by better concepts. For this reason, it is difficult for any science publication to be completely comprehensive. It is possible, however, to provide a thorough foundation that ensures that any such advances can be comprehended. It is the purpose of each book in this series to create such a foundation, by providing all the basic knowledge in the particular area of science it describes.

How to use this book

This book can be used in two basic ways.

The first, and more conventional, way is to start at the beginning and to read through to the end, which gives a coherent and thorough picture of the subject and opens a resource of basic information that can be returned to for re-reading and reference.

The second allows the book to be used as a library of information presented subject by subject, which the reader can consult piece by piece as required.

All articles are prepared and presented so that the subject is equally accessible by either method. Topics are arranged in a logical sequence, outlined in the contents list. The index allows access to more specific points.

Within an article, scientific terms are explained in the main text where an understanding of them is central to the understanding of the subject as a whole. There is also an alphabetical glossary of terms at the end of the book, so that the reader's memory can be refreshed and so that the book can be used for quick reference whenever necessary.

Each volume also contains a section on the various careers that pertain to the volume's subject.

The sample two-page article *(right)* shows the important elements of this editorial plan and illustrates the way in which this organization permits maximum flexibility of use.

(A) **Article title** gives the reader an immediate reference point.

(B) **Section title** quickly shows the reader how information is arranged within the article.

(C) **Main text** consists of narrative information set out in a logical manner, avoiding biographical and technical details that might tend to interrupt the story line and hamper the reader's progress.

(D) **Illustrations** include specially commissioned drawings and diagrams and carefully selected photographs, which expand, clarify, and add to the main text.

(E) **Captions** explain the illustrations and make the connection between the textual and the visual elements of the article.

(F) **Labels** help the reader to identify the parts of the illustrations that are referred to in the captions.

(G) **Theme images,** where appropriate, are included in the top left-hand corner of the left-hand page, to emphasize a central element of information or to create a visual link between different but related articles.

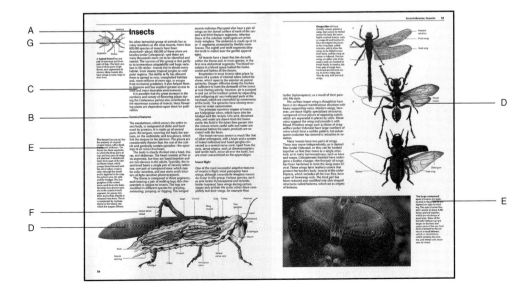

Entomologists collect insects in a canyon pool in Mexico. Insect collection and other types of field work are an important part of an entomologist's job.

Entomologists study insects and related animals, such as ticks, mites, spiders, and centipedes. Most entomologists work in the field of *applied entomology*, which focuses on insect pests that damage crops and other plants, products in storage, and even buildings. These scientists also study insect pests that endanger the health of people and animals. In addition, they investigate ways to control insect pests. Entomologists use a variety of controls to reduce insect populations. These include cultural controls, such as the draining of swamps where mosquitoes breed; chemical controls, such as the use of insecticides; and biological controls, such as the introduction of animals that naturally prey on insect pests. Entomologists also look for ways to protect insects that are helpful to people, such as silkworms and honey bees.

Some entomologists work for state and federal agriculture experiment stations, public health agencies, and companies that manufacture pesticides. Others teach and conduct research at agricultural universities.

Ethologists are zoologists who study animal instincts to learn what causes instinctive behavior, how it has evolved over time, and how it helps a species survive. Ethologists observe an animal's normal activities in the wild to learn more about the wide range of instinctive behavior in animals, including courtship, mating, and caring for young. In addition, ethologists study how animals communicate with each other and how they establish and defend their territories. For every animal they study, these scientists develop an *ethogram*—a list of all the known behavior patterns of a species, including the conditions under which each instinctive act occurs.

Herpetologists are zoologists who specialize in the study of reptiles and amphibians, including snakes, lizards, crocodiles, frogs, and salamanders. Herpetologists study the evolution and behavior of these animals and observe how reptiles and amphibians interact with other animals and with their environment. Herpetological research has made many significant contributions to the field of biology, particularly our understanding of ecology, evolution, and animal behavior.

Today, the future of many wild reptiles is threatened, so herpetologists often assist government agencies and conservation groups in developing ways to protect the natural habitat of these animals. They also educate people about the important role of reptiles and amphibians in the ecological cycle, and how hunting them and collecting their eggs is endangering many species.

Ichthyologists are zoologists who specialize in the study of fish, including their development, structure, and behavior. They also investigate the geographical distribution and migratory habits of fish, and how fish relate to their environment.

The research conducted by ichthyologists is important in many ways. For example, ichthyologists help commercial fisheries and government conservation agencies manage the world's vital fishing areas. Their work helps ensure that these areas are not overfished or threatened by pollution. Ichthyologists also work with *aquaculturists*, who raise marine animals in fish farms for food production, to help them improve their crops and prevent disease among the fishes.

Ornithologists specialize in the study of birds. They observe birds in the wild to learn more about their behavior and their numbers as well as their history, physical characteristics, and geographic distribution. Ornithologists often use photography to document how birds breed, nest, feed, navigate during flight, and migrate. They also make recordings of birdsongs to learn more about how birds communicate.

Some ornithologists work in museums and zoos where they manage bird collections, conduct research, and prepare exhibits and educational publications. Others conduct bird population counts for government agencies and help conservation officials determine which species need special protection.

Veterinarians diagnose, treat, and prevent illness in large and small animals. In cities, veterinarians work in animal hospitals, caring for our pets—dogs, cats, and other creatures. Today's veterinarians use medical equipment and techniques that are similar to those used for humans. In rural areas, most veterinarians focus their efforts on the care and treatment of farm animals, using their special skills and knowledge to prevent the outbreak of disease, which could kill an entire herd. Veterinarians employed by government and public health services work to control the spread of animal diseases that can be transmitted to humans. To become a veterinarian, a student must first complete at least two years of preveterinary college work and then study in a college of veterinary medicine for four years.

A veterinarian operates on a dog with the aid of a veterinary technician. Veterinarians often use medical equipment and techniques that are similar to those used for humans. Their technicians assist them in a wide range of activities, including record keeping, lab work, and surgery.

Veterinary technicians work with veterinarians in caring for animals much as nurses assist doctors in caring for people. These technicians often work in animal hospitals, where they perform a wide range of duties. They record information on patients, collect specimens, and perform laboratory tests. They also prepare patients and equipment for surgery, assist the surgeon, and check on recovering patients. Veterinary technicians are concerned with an animal's emotional needs as well as physical well-being. They provide comfort and reassurance to anxious animals—and their owners. They also instruct pet owners in the proper care of their animals, such as giving medication, changing bandages, and watching for changes in the animal's condition. These technicians receive their training in a two-year, college-level veterinary technician program.

Wildlife biologists study the geographical distribution, habitat, ecology, mortality, and economic implications of animals in the wild. Some wildlife biologists treat sick and injured animals at animal rehabilitation centers and, whenever possible, return them to their natural home. Some wildlife biologists conduct research on fish and wildlife populations to determine their environmental and nutritional needs and their relationship with other animals and plants. This research is used to develop wildlife programs, create or restore habitats, and help control disease in wildlife.

Wildlife photographers take pictures of animals in their natural habitat for books, television and videotape programs, and movies. Most of a wildlife photographer's work is intended to educate the viewer about the subject, so their films and pictures must depict animals realistically, and they must be knowledgeable about the behavior and routines of the animals they photograph. Wildlife photographers often travel to remote places to capture an unusual species on film for a magazine or documentary film. Some wildlife photographers also conduct photography expeditions, or "photo safaris," in which they take small groups of photo enthusiasts into areas with abundant wildlife, such as Africa.

King penguins pose for a wildlife photographer on a snow-packed field on South Georgia Island. Wildlife photographers must often travel to remote places and work under difficult conditions to capture wild animals on film.

Zoo biologists use medical technology to breed endangered or threatened animals in the controlled environment of a zoological park. They conduct research in reproductive physiology to help animals who have difficulty reproducing in captivity as well as animals who are not breeding successfully in nature. In addition, zoo biologists conduct observational research to find out how animals reproduce under natural conditions, and perform laboratory tests to learn which animals are best suited for breeding. They also collect sperm and eggs from living animals for use in a technique called *in vitro fertilization*. In this technique, an embryo is created by combining a sperm cell and egg cell in a laboratory dish. The embryo is then surgically implanted in the body of a female animal. Many animal species are now in danger of extinction in their natural habitat, so the work of zoo biologists is vitally important.

Zookeepers are responsible for the care and maintenance of animals living in zoos. Their daily tasks include feeding the animals and cleaning their living quarters. Zookeepers work closely with veterinarians to ensure the health and well-being of the animals in their care. Many exotic animals live in zoos, and zookeepers must be keenly aware of their special needs. They watch the animals closely for any change in behavior that may indicate illness or disease. Because there are so many different kinds of animals in zoos—and so much to know about them—many zookeepers work with a single animal species. In addition to caring for the animals, zookeepers are often asked to share their knowledge of an animal's habits with visitors and to communicate our obligation to protect animal species from extinction.

Zoologists study animals and their behavior. These scientists observe how animals interact with other animals, with their environment, and with people. They also conduct research to learn how animals have evolved over time. Many zoologists work in laboratories, zoos, and museums. Others travel to faraway places to observe animals in their natural environment. These expeditions are called field studies.

Because there are so many animals—and so much to learn about them—most zoologists specialize in a particular area. Many branches of zoology focus on a specific species. Other specialties include *taxonomy*—the naming and classifying of animals—and *paleontology*—the study of prehistoric animals. Some zoologists specialize in wildlife management or in domestic animal breeding programs. Because animals and humans are similar in many ways, zoological studies help us understand more about human disease. The work of zoologists benefits both animals and humans.

A zookeeper tends a giant tortoise, one of the many exotic animals that live in zoos. Such exotic animals have special needs and a zookeeper must attend to these needs to ensure the animal's health and well-being.

Principles of zoology

Major groupings of animal classification range from phylum to species. The system can be illustrated by looking at the complete classification of the African lion:

Phylum
Chordata
Subphylum
Vertebrata
Class
Mammalia
Order
Carnivora
Family
Felidae
Genus
Panthera
Species
P. leo

The lion is uniquely identified by its binomial Latin name, *Panthera leo*.

There are almost 1 million different species of animals living in the world today, and new ones are constantly being discovered. They range in size and complexity from microscopic creatures made up of only a few cells to the giant blue whale, weighing up to 180 tons. Faced with the task of bringing order to the living world, zoologists consider two key factors: definition (is it an animal?) and classification (what place does it occupy in the animal kingdom?).

What is an animal?

Apart from inanimate objects, humans share the world with living organisms, including plants, animals, and single-celled organisms called monerans and protists. The difference between inanimate objects and living things is not always apparent. For example, one of the characteristics of living things is that they grow. Yet crystals, which are inanimate, also grow. But only living things share in common the fundamental life process called metabolism. Metabolism is the sum of the chemical processes by which cells produce the substances and energy that an organism needs to live.

Almost all the world's energy is provided initially by the sun. Plants trap a proportion of this energy and use it, through photosynthesis, to produce the sugars and starches that form a major part of their tissues. During this process they absorb minerals and water from the soil and carbon dioxide from the atmosphere. They release oxygen, which is a waste product of their metabolism.

Animals, however, are more complex—they cannot use such simple chemicals to produce the substances from which they are made. Instead, they must eat and digest food, and reorganize the products of digestion to form their own tissues. Animals require oxygen to release the potential energy of their food, and they emit carbon dioxide as a waste product.

Thus, plants and animals are complementary, each using the other's wastes, which are constantly recycled. This great continuous energy flow is aided by a variety of decomposers, especially bacteria and fungi. Decomposers break down dead material and waste into a form that plants can use again.

Animals differ from plants chiefly by their quick responses to stimuli and their ability to move about easily. The cells of plants and animals differ greatly, too. Only plants contain the pigment chlorophyll and have a substance called cellulose in their cell walls.

The distinction between a plant and an animal is often difficult to make among the simplest organisms, however. In the past, scientists classified the simple, single-celled organisms called protozoans as animals because they hunt and feed. But some of these organisms also contain chlorophyll, which they use to make their own food. As a result, today most scientists classify protozoans in a kingdom separate from plants and animals.

Animal species

Zoologists generally classify animals according to a system devised by the 18th-century botanist, Carolus Linnaeus. In this system, animals are divided into ever-smaller groups that have more and more features in common. The largest groups are the kingdoms. All animals belong to the kingdom Animalia. The next largest group is a phylum. Each phylum is divided into classes, classes into orders, and orders into families. Families are then broken down into genera (singular, *genus*) and genera into species, the basic unit of biological classification.

Animals of the same species are similar in structure and function and can breed with one another freely in the wild. Animals of different species, though they may resemble one another, generally cannot produce offspring together. The words "freely" and "in the wild" are important qualifications because people have successfully bred different species of animals together. A cross between a horse and a donkey, for example, produces a mule. In these kinds of crosses, however, the resulting offspring, called a hybrid, is usually completely sterile or lacks the vigor necessary to compete with other animals for mates.

However, scientists agree that within species, there is room for genetic diversity.

They also agree that species change gradually, adapting to environmental variations as successful genetic strains outbreed the less successful ones.

Classification

Scientists need an internationally accepted classification system for animals for two main reasons: first, to be able to identify a particular species throughout the world, and second, to group related kinds into larger groups that reflect evolutionary relationships. Common names are usually too imprecise for scientific use. The system of scientific classification for animals includes the categories species, genus, family, order, class, subphylum, and phylum. Using these categories, it is possible to see the relationship between animals of different species at each level.

The lion, for example, is designated by the scientific name *Panthera leo,* which indicates that it is the species *leo* belonging to the genus *Panthera.* This genus also includes the jaguar *(P. onca)* and the tiger *(P. tigris).* These animals all belong to the family Felidae (the catlike animals), which belongs to the order of flesh-eating animals called Carnivora. This order includes other animals with similar tooth arrangement, skull form, and other features.

As a member of the order Carnivora, the lion does not much resemble the anteater, for instance, which belongs to the order Edentata ("toothless"). But both have mammalian features and so are designated members of the class Mammalia. All mammals have a backbone and are therefore placed in the subphylum Vertebrata, along with fishes, amphibians, reptiles, and birds. Vertebrates and other animals with a notochord, which is a stiff internal supporting rod (in vertebrates, present only in the unborn and larval states), are placed in the phylum Chordata.

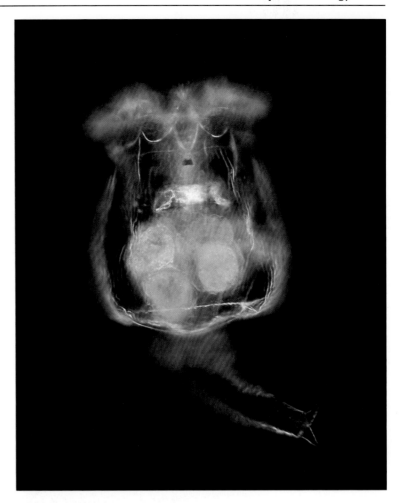

Rotifers are named after their wheellike corona, which beats in a circular motion to achieve movement. These aquatic animals occur in abundance—about 1,500 species are known to exist—and are among the smallest multicellular organisms.

Greater and lesser flamingos inhabit lakes in tropical Africa. Flamingos belong to the class Aves, or birds. Like all birds, flamingoes breathe air by means of lungs; lay eggs; and have a toothless beak, two legs with scaly feet, and, most importantly, feathers. The greater flamingo sifts worms, mollusks, and small arthropods from the mud, whereas the lesser flamingo filters blue-green algae from the lake bottom. These birds thrive in this environment because their long legs and necks make it easy for them to wade and feed in deep water.

The animal cell

The term "cell" was first used by the English naturalist Robert Hooke in 1665 to describe the "great many little boxes" he saw when viewing a thin slice of cork through a microscope. He derived the word from the Latin *cella,* meaning a small room. Hooke was the first scientist to recognize that living matter was built of basic units rather than of continuous material.

Today, with the aid of electron microscopes, we have a more detailed view of living plant and animal cells. Some plant and animal cells have a particular function and are therefore not all exactly the same. The cells of animals exist in a wide variety of shapes. They may be round, egg-shaped, square, or rectangular. Some muscle cells are long and thin, and pointed at each end. Some nerve cells, with their long branches, resemble trees.

A system of similar cells forms a tissue—for example, nerve cells make up nerve tissue—and combinations of tissues in turn form an organ, such as the brain.

The structure of cells

Despite this variety of form, all cells have some features in common. A cell consists of a nucleus, which is surrounded by a jellylike substance called the cytoplasm, enclosed in a cell membrane. This cell membrane (also called the plasma membrane) controls the passage of substances into and out of the cell. The membrane, only 0.00001 millimeter thick, is made up of layers of lipid, or fat, molecules sandwiched between layers of protein. The cell membrane is semipermeable—that is, it allows certain chemicals to move in and out of the cell and prevents the passage of others.

The cytoplasm contains several kinds of structures, among which are many tiny structures called organelles. Many of the cell's life activities take place in the cytoplasm. Each organelle plays a vital role in maintaining the life of the cell.

Organelles

Most of the cytoplasm of a mature cell is filled with elaborate folded membrane systems called the endoplasmic reticulum. The cytoplasm and the soluble proteins of the cell lie on one side of the endoplasmic membranes. On the other side, many cells contain enclosed pockets called cisternae. It seems likely that the endoplasmic reticulum plays an important role in the transport of substances throughout the cell. It also provides a large surface area where essential chemical reactions take place. Many enzymes that are important for the cell's metabolism are found on the cytoplasmic side, whereas their products are found in the cisternae.

There are two types of endoplasmic reticulum. Smooth endoplasmic reticulum helps manufacture fat molecules. Rough endoplasmic reticulum is involved with the synthesis of proteins. Attached to the walls of the rough endoplasmic membranes are granular structures called ribosomes. But not all of the cell's ribosomes are attached—some float freely in the cytoplasm. Those that are attached to the endoplasmic reticulum help build proteins that are transmitted to other parts of the body. Those that float freely aid the synthesis of the proteins that remain in the cell.

The rough endoplasmic reticulum also produces packages of enzymes called lysosomes. Under acidic conditions, lysosomes rupture and release their enzymes, which break down the major components of the cell. Lysosomes also release enzymes that destroy damaged or

An animal cell (below) is characterized by the various structures within it, chief of which are the nucleus (with its nucleolus), mitochondria, and the folded layers of the endoplasmic reticulum.

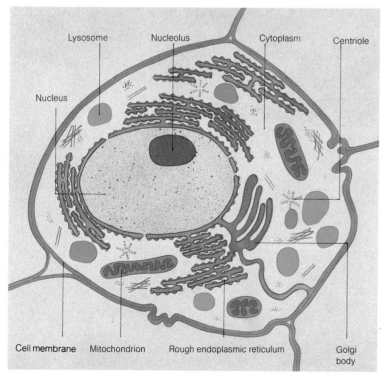

Lysosome Nucleolus Cytoplasm Centriole

Nucleus

Cell membrane   Mitochondrion   Rough endoplasmic reticulum   Golgi body

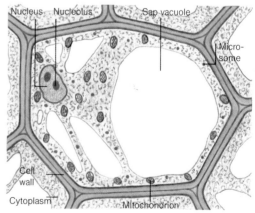

Nucleus Nucleolus Sap vacuole

Microsome

Cell wall

Cytoplasm Mitochondrion

A typical plant cell *(above)* differs from an animal cell in that it has rigid cell walls, a large permanent vacuole in the center of the cytoplasm; and chlorophyll, located in the chloroplasts.

unneeded cells and that digest "foreign" cells, such as bacteria.

The Golgi complex consists of a stack of flat membrane sacs. These sacs process proteins and other substances produced in the cell. Small spheres called vesicles pinch off from the Golgi complex and transport some of these substances across the membrane to other cells or they are used to make the cell's covering. Other Golgi vesicles remain inside the cell and fuse together to form storage compartments for proteins or other substances.

Mitochondria, another group of organelles found in the cytoplasm, are the powerhouses of the cell. They generate the energy that is needed to keep the cell's essential processes going. Mitochondria are elongated, fluid-filled structures enclosed by a double membrane. The inner membrane comprises a series of deep folds called cristae. The enzymes needed to extract energy are located on the inner membrane. They oxidize nutrients and release energy in the form of a compound called adenosine triphosphate (ATP), which is used in the synthesis of cell materials in a process called internal respiration, or cellular respiration.

Centrioles, another type of organelle, are important in cell reproduction. They lie near the nucleus and resemble two bundles of rods.

The nucleus

The nucleus controls all the cell's activities. It is spherical or elliptical in shape and bounded by two membranes that together form the nuclear membrane. The outer membrane, which seems to be an extension of the rough endoplasmic reticulum, has several small pores through which nuclear material and large molecules pass.

A substance called the nucleoplasm inside the nuclear membrane contains the nucleolus and chromosomes. The nucleolus, a spherical body made up of ribonucleic acid (RNA), helps in the formation of ribosomes, which are the cell's centers of protein production. RNA is chemically similar to DNA and plays an important role in protein synthesis. Most cells contain one or more nucleoli. The chromosomes, composed chiefly of deoxyribonucleic acid (DNA), are the blueprints for the cell's structure in the form of a genetic code. In the cell's resting phase, the DNA is distributed throughout the nucleoplasm in the form of a threadlike material called chromatin. When a cell is dividing, the fine threads of chromatin shorten and thicken to form visible chromosomes.

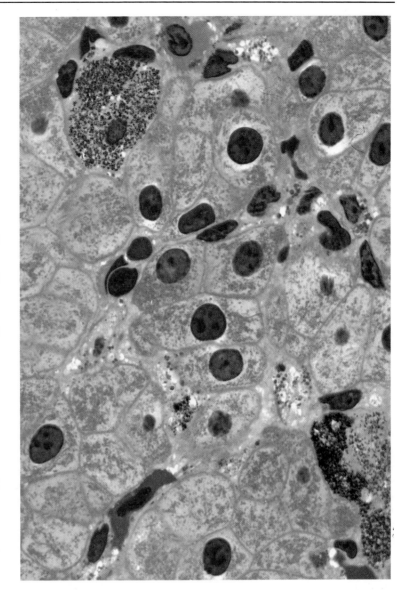

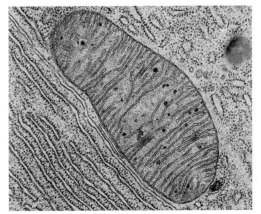

Liver cells from a salamander are stained *(above)* so that they can more easily be studied. They would otherwise be impossible to see.

All animal cells have at least one mitochondrion *(left)* and usually many more. These powerhouses are the site of the energy-forming processes of the cell.

Anatomy and physiology

All organisms, whether plant or animal, must reproduce, grow, breathe, eat, excrete, and be able to respond to their environment. Various systems in animals' bodies accomplish these tasks. These systems may vary from species to species, but the chemical processes involved in the activities are usually the same.

Differences in anatomy, or arrangement of limbs and organs, have arisen for several reasons. One is that for early aquatic animals to evolve into more advanced forms that could live on land, various physical features had to be modified. For example, legs and feet replaced fins; and lungs, which enabled animals to get oxygen directly from air, replaced gills.

One of the most important anatomical developments in animals is segmentation, first seen in the earthworm, in which the body is divided into ringlike segments. The more primitive animal phyla lack segmentation, which occurs in all higher groups. In the more advanced animals, such as vertebrates, segmentation occurs only in the embryo stage.

Structural variations

Three basic types of support systems, or skeletons, exist in the animal kingdom: the hydrostatic skeleton, the exoskeleton, and the endoskeleton. Some invertebrates, such as jellyfish, worms, and sea stars, have a skeleton composed of a simple, fluid-filled cell called a hydrostatic skeleton. These animals move by contracting muscles, which shift water held inside the body, bringing about a change in the animal's shape or position.

The greater gliding possum glides from one treetop to another on a membrane that stretches from its neck to its feet. This structural adaptation to its heavily forested environment enables the animal to get around without having to run down one tree and climb up another.

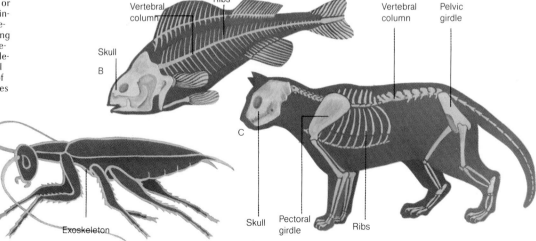

The supporting structures of animals consist of the hydrostatic skeleton or fluid-filled cell of lower invertebrates; the exoskeleton or hard outer covering (A) found in other invertebrates; and the endoskeleton, which is the internal skeleton characteristic of vertebrates, such as fishes (B) and cats (C).

Other invertebrates, such as insects and crustaceans, have a hard outer case shell called an exoskeleton. The exoskeleton, which forms from secretions of the animal's skin, is segmented and bears jointed legs. The muscles that enable these animals to move are attached to the inside of the skeleton.

Vertebrates have an endoskeleton, which is a hard internal support system made of bone. Muscles fixed to the end of each bone work in pairs, one contracting while the other relaxes, enabling vertebrates to move. The joints of the vertebrate body are enclosed in a synovial capsule and are lubricated by synovial fluid.

The basic structure of an animal also determines how it will grow, and the growth patterns of animals differ greatly. Because an exoskeleton does not grow, insects and crustaceans must molt, or shed their exoskeleton periodically, as they outgrow it. Mammals have special growth centers near the tips of their long bones, where the cell division that enables growth takes place. In addition, mammals' skull bones are separated by cartilage, which allows the skull to grow. Growth stops when the bones meet.

Animals also display a wide variety of adaptations that enable them to move about. Mammals, birds, and many reptiles and amphibians have legs with feet that enable them to walk on land. Other animals can crawl without legs. Some invertebrates, such as planarians and flatworms, slide along by moving tiny hairlike structures called cilia. Snails coat the ground with a sticky fluid and then glide along on a muscular organ called a foot. Snakes use their muscles to bend their bodies back and forth, enabling them to slither along the ground.

Animals also have ways to get around in the air and water. Insects, birds, and bats have wings that allow them to fly. Many animals that live in water, such as fish and whales, have well-developed fins and tails that they use to propel them through the water. Other aquatic animals, such as jellyfish and squid, swim by jet propulsion. Many birds can dive and swim by using their feet and wings as paddles or oars.

Respiration

The chemical processes that result from gas exchange are necessary to the survival of most animals. These processes take place within the respiratory system. Respiratory organs take a variety of forms. But from the simplest to the most complex, they are mainly concerned with two processes: external respiration, which involves the intake of oxygen and its transport to each cell in the body, and internal respiration, the chemical processes that use the oxygen.

External respiration finishes when the oxygen crosses the respiratory surface, the place where the exchange of gases takes place, and enters the cells. The respiratory surface varies from animal to animal, but it has certain features in common: it is moist and thin (normally one cell thick); it is permeable, allowing substances in solution to pass through it; and in some animals, it is supplied with blood vessels or some other means of transporting oxygen to the cells. Oxygen enters jellyfish and related animals through the surface cells of the entire animal, so that the whole surface of the animal is its respiratory surface.

The moist skin of the earthworm also acts as the respiratory surface, but its body is too thick for the oxygen simply to diffuse from the outer layer to the inner cells. To transport dissolved oxygen to all parts of the body, the earthworm has a blood circulation system.

In land-dwelling arthropods, such as scorpions, spiders, and insects, a series of air tubes called tracheae runs from the outer surface into the body and ends in special areas of gas exchange called alveoli. Each alveolus is kept moist so that oxygen can dissolve into the water and thus leave the alveolus to diffuse into the body cells. Aquatic invertebrates with an exoskeleton, such as crabs and lobsters, and aquatic vertebrates, such as fishes, have a gill system that acts as a respiratory surface, extracting oxygen from the water.

The lungs of land vertebrates contain air tubes, alveoli, and a supply of blood vessels to carry the oxygen to other parts of the body. In mammals, a ribcage protects the lungs, and a muscular sheet called the diaphragm separates them from the rest of the internal organs. The ribs and diaphragm act together to pump a continuous stream of fresh air into the lungs. Air entering the lungs dissolves in the film of water in the alveolus chamber and diffuses through the chamber wall into the blood vessels. A red substance called hemoglobin in the blood transports the oxygen to the cells.

Digestion. Insects (A) have a mouth and a simple digestive tract. But vertebrates (B and C) have a higher metabolic rate and so need more energy. They therefore have more efficient and specialized digestive systems that can extract high amounts of energy from food.

Fledgling songbirds are nestbound and unable to see at birth, and for several weeks are dependent on their parents for food. In contrast, many less sophisticated animals, such as larvae and young insects, are able to fend for themselves immediately after birth. The vulnerability of young birds makes them easy prey for predators and means that they need constantly to be guarded by their parents.

The nervous system

Every animal has a nervous system, which allows it to sense and interact with its environment. Jellyfish and sea anemones have identifiable nerve cells that form a nerve net across the surface of their bodies, but they lack a central nervous system and respond generally to stimuli such as light and dark and hot and cold.

In worms and arthropods, some nervous tissue is concentrated at the front of the body. In these animals, a primitive brain joins a thick ventral nerve cord that runs the length of the body. Smaller nerves branch from it to other parts of the body, and so a basic information-collecting system is formed.

Vertebrates have a specialized complex nervous system composed of a brain and a spinal cord, which is protected by the spinal column. The brain is connected to the eyes, nose, ears, and other parts of the head by a series of cranial nerves. Spinal nerves connect the rest of the body to the spinal cord. The warm blood of mammals and birds provides the stable internal environment necessary for the development of their large brains.

The eyes of animals provide evidence of the evolution of sensory organs. The eyes of some crustaceans and worms are sensitive only to light—these animals have simple clusters of photoreceptors, called ocelli, which guide them toward and away from light. In higher animals, the sensory cells are so organized that an image forms inside the eye. Lower vertebrates such as reptiles, amphibians, and fish have lateral vision, in which each eye—one on each side of the head—forms a separate two-dimensional image of a scene. But mammals and birds of prey have three-dimensional stereoscopic vision. Their eyes are directed forward so that a single image of a scene is formed.

Chewing and digestion

Most animals merely swallow their food, making no attempt to chew it. In some groups, such as birds and some insectivores, tiny stones in the stomach help break the food down so that enzymes can digest it further. Most vertebrates use their teeth only to hold food before swallowing it whole.

Mammals, however, must begin the digestion process in the mouth by chewing. This helps them get the maximum nutritional value from the food they eat to supply the increased energy needed for warm-bloodedness. Mammals have developed different types of teeth depending on their diet. Herbivorous, or plant-eating, animals generally have long jaws with a set of flat-topped, grinding cheek teeth to break up tough plant material. Carnivorous, or meat-eating, animals have sharp teeth, or canines, which they use to kill their prey, along with slicing cheek teeth, or carnassials, which they use to cut the meat into swallowable pieces. Omnivorous mammals, which eat both meat and plants, have a combination of flat-topped cheek teeth with sharp, cutting edges and incisors for biting.

In all animals, whether simple or complex, enzymes convert food to its simplest form. This process is called digestion. Jellyfish, sea anemones, and their relatives take food in through the mouth—the only opening in the body. The lining of the internal cavity then secretes enzymes that digest the food material, and special cells inside the cavity engulf the

Circulation. Insects (A) have a single blood vessel which widens into a 13-chambered, tube-shaped heart in the abdomen. Blood is not restricted to this vessel, but flows throughout the body cavity through spaces called blood sinuses. Muscular contractions in the vessel pump the colorless blood forward through the blood vessel, and it seeps through the blood sinuses to reenter the heart. Fishes (B) have a ventral blood vessel running along the underside of the body, from which segmented vessels branch and run through the gills to join the dorsal vessel, which runs along the top side of the body and distributes blood to the rest of the fish's body. Mammals (C) have a four-chambered heart. Oxygenated blood is distributed by arteries, and veins return deoxygenated blood to the heart.

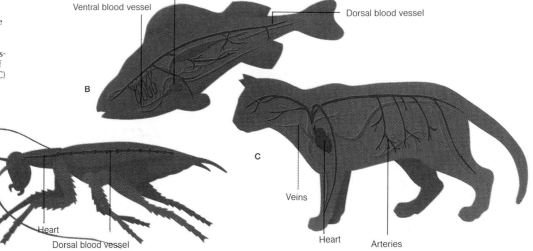

digested food. Unused food is rejected from the body via the mouth.

Most other animals have a tube called the alimentary canal that runs from the mouth to an opening called the anus in the rear. The food enters the tube via the mouth, which opens into the buccal cavity, where the digestive process begins. The food then passes along a muscular section of the tube, pushed by wavelike contractions called peristaltic waves, until it reaches the stomach. Most of the digestive process occurs in the stomach or an area close to it, such as the intestines. Once digestion is completed, the body absorbs the useful materials, and the waste products are passed out of the body via the anus.

In addition, vertebrates and some other animals have specialized glands that produce secretions that aid the digestive process. In mammals, for example, the liver produces bile, and the pancreas releases digestive enzymes into the small intestine. Food broken down by enzymes is absorbed into the blood and taken to the liver, which begins the process of eliminating any unnecessary chemicals.

Most animals also contain beneficial microorganisms that live in their digestive systems. Some animals carry on symbiotic relationships, in which another organism lives within or on the body of another for the mutual benefit of both. For example, single-celled organisms called protozoa live in the guts of some termites and break down the wood that they eat.

Excretion

Excretion is the process by which animals eliminate waste from their bodies. In many simple organisms waste passes out via the body wall. In larger animals, however, special organs carry on this process.

The circulatory and respiratory systems in vertebrates remove carbon dioxide and water from the body, and the kidneys filter out excess chemical substances brought to them by the blood. The toxic substance ammonia is the end product of metabolism. A great deal of water is required to flush ammonia out of the body. Aquatic animals such as fishes, for which water loss is not as great a problem, excrete ammonia directly. Land animals convert ammonia into less toxic urea and uric acid, which are then collected in the bladder, mixed with a smaller amount of water, and are excreted in the form of urine. Some animals, such as insects and land snails, which need to stay moist by keeping water in their bodies, excrete dry uric acid crystals.

The newborn young of some grazing animals are fully active soon after birth—this young wildebeest could get on its feet within seven minutes of being born. They feed on mother's milk for several months but, unlike nest-bound fledglings, can move around independently.

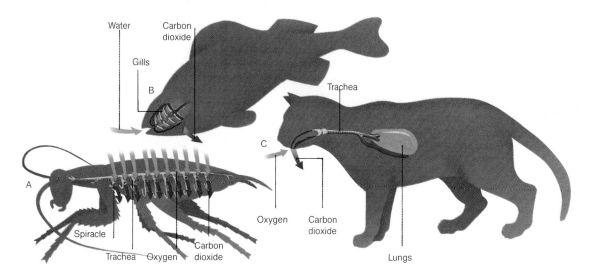

Respiration. In insects (A), air is conducted through openings called tracheae to the blood, into which oxygen diffuses. Fishes (B) take water in through the mouth; the water passes over the gills, which extract oxygen from it. Mammals (C) breathe with lungs, where oxygen diffuses into blood vessels.

Water

Carbon dioxide

Gills

B

Trachea

C

A

Oxygen

Carbon dioxide

Spiracle

Trachea   Oxygen   Carbon dioxide

Lungs

Reproduction

Death from old age, predation, disease, injury, or sudden environmental change is an ever-present threat to animals. A species can survive and adapt, however, because its members reproduce themselves. When animals reach adult size, they become sexually mature and able to reproduce. They do so in one of two basic ways—by asexual or sexual reproduction. Asexual reproduction involves only one parent, whereas sexual reproduction usually involves two adult individuals.

In many animals chemicals called hormones and pheromones cause the body to develop reproductive cells. The production of hormones also influences courtship and mating behavior. In certain climates, environmental conditions, such as the longer days of spring and early summer may activate hormones and pheromones.

Asexual reproduction

Only the more primitive animals reproduce asexually. It is a less complicated means of reproduction and has the disadvantage of producing offspring that are identical to their parents. There is no shuffling of genetic material between generations and therefore less variation within the population.

The simplest method of asexual reproduction—binary fission—is found in unicellular organisms such as amebas. These organisms simply split their nucleus in two by mitosis (cell division) and the cytoplasm separates to surround each new nucleus. The parent no longer exists as a single unit but has become two "daughter" cells.

Some animals that are capable of sexual reproduction may also reproduce by means of two other methods of asexual reproduction—fragmentation (or regeneration) and budding. Regeneration is the process by which a new adult individual grows from a body part that has broken off. Hydras and sponges reproduce through regeneration. In budding, an outgrowth, or bud, develops on the body. The bud then grows into a smaller copy of the parent and eventually detaches itself. In certain corals, for example, buds appear on the body

Budding is a method of asexual reproduction characteristic of cnidarians such as the hydra. Budding differs from regeneration in that specialized cells are involved. These cells on the body surface grow rapidly to form an outgrowth, or bud, and develop into a smaller copy of the parent.

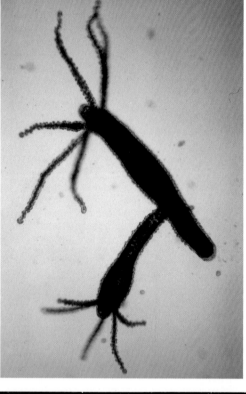

During spawning, the female Surinam toad lays a few eggs at a time, which the male fertilizes and then presses into the softened skin on the female's back. The mother's skin forms a small capsule for each of the eggs, and the young hatch and pass the tadpole stage in her back. Fully developed toads emerge from the mother's skin when they are about $2\frac{1}{2}$ months old.

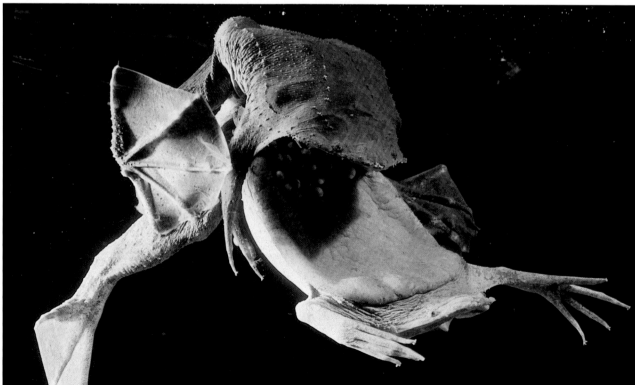

of an adult. These buds grow larger, separate from the parent, and begin to deposit their own limestone in the colony. In this way, thousands of individual animals, all derived from a single parent, may form great colonies of coral. Members of some animal groups, such as cnidarians, produce young both sexually and asexually, alternating generation by generation. Usually one stage is the dominant state for a species, while the other is a temporary phase.

Sexual reproduction

The two parent organisms involved in sexual reproduction each produce special sex cells called gametes. Female gametes, called ova or eggs, are formed in an ovary. They are usually much larger than the male gametes, called sperm, and contain nutrients that feed the embryo. Ova are the largest single cells and have no means of locomotion. Sperm are formed in the testicles and move by means of a whiplike tail called a flagellum.

Gametes contain half the number of chromosomes that every other adult cell in the body possesses and are called haploid, whereas the body cells contain paired sets of chromosomes (homologous chromosomes) and are called diploid. The reduced number of chromosomes in gametes results from a special kind of cell division called meiosis.

When the nuclei of two gametes join during fertilization, the resulting cell, or zygote, has the full number of chromosomes and two sets of hereditary information, one from each parent. The offspring exhibit features of both parents but are not exact copies of either.

Mating

Many animals exhibit specific courtship behavior to help them find suitable mates. This behavior tends to follow a particular pattern according to species, ensuring that animals mate only with members of their own species. Animal mates find each other in a number of ways. Female birds are attracted to the songs and brightly colored feathers of male birds. Many male insects and amphibians attract females with their loud calls. Some animals give off scented chemicals called pheromones to attract the opposite sex.

Other animals perform ritualized movements to attract mates. Male fence lizards, for example, bob their heads rhythmically when a female approaches. Still others offer food to possible mates. A male tern catches a fish and places it into the mouth of the female he wants to mate.

Mating is dangerous for some male spiders and insects. Male black widow spiders and male praying mantises, for example, are sometimes eaten by females after mating.

Fertilization

In most animals the male and female sex organs are found in different individuals. However, in some animals, such as hydras, flatworms, earthworms, and snails, are hermaphrodites, in which both ovaries and testes are present in each individual. Tapeworms fertilize themselves, but most hermaphrodites cross-fertilize with other members of their species.

Mitosis (A) is the process by which all body cells are produced. Meiosis (B) occurs only in sex cells. Both processes involve several phases during which a cell containing a complete set of chromosomes is split. During prophase, the centrioles, bundles of DNA-containing rods outside the cell's nucleus, separate and form a spindle around the nucleolus, which disap-

pears. Each chromosome consists of a pair of identical chromatids jointed by a centromere. The chromosomes collect at the spindle's equator with the centromeres on the spindle threads. In mitosis, the centromeres divide and migrate to opposite poles, but in meiosis whole chromosomes migrate. The cell then splits. In mitosis the chromosomes unwind and

revert to chromatin threads. Each daughter cell contains one half of each original pair of chromosomes and replicates the other half. In meiosis, the two new cells split again so that the four resulting cells are haploid, containing only one half of each original pair of chromosomes.

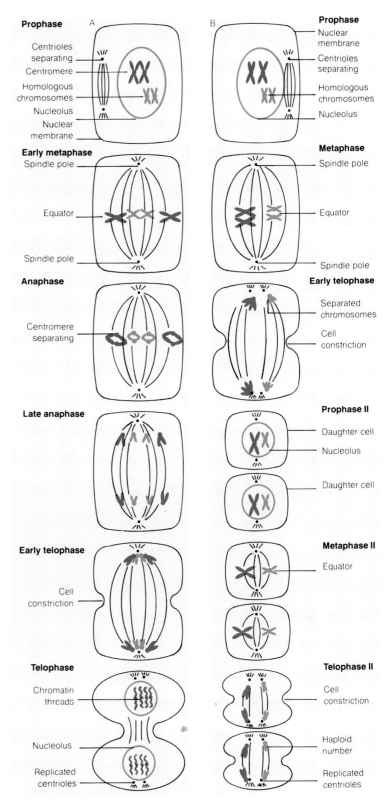

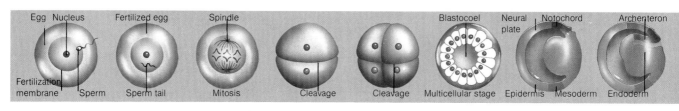

Egg Nucleus   Fertilized egg   Spindle   Blastocoel   Neural plate   Notochord   Archenteron

Fertilization membrane   Sperm   Sperm tail   Mitosis   Cleavage   Cleavage   Multicellular stage   Epidermis   Mesoderm   Endoderm

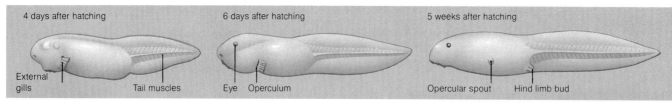

4 days after hatching   6 days after hatching   5 weeks after hatching

External gills   Tail muscles   Eye   Operculum   Opercular spout   Hind limb bud

The fertilization of a frog egg *(top)* is followed by mitotic division of the egg, called cleavage. The egg cleaves into a two-, four-, and then eight-celled organism. A cavity (blastocoel) forms as cleavage continues and different types of cells that make up various tissues begin to form.

Young mammals depend on their mother for food, which, during the first few weeks or months of life, takes the form of milk. Parental care also includes keeping the infants clean and protecting them from predators.

For fertilization to occur, it is important that the sperm and eggs are deposited close to one another. The mobile sperm swim to an egg, attracted by chemicals in the surrounding fluid; but because sperm consist mostly of genetic material, they have a very small energy store and usually cannot survive for more than 24 hours under normal conditions. For this reason, fertilization can take place only in liquid. Animals that live in water, such as fishes and frogs, can reproduce by means of spawning because their reproductive cells are already surrounded by water. In spawning, the female usually releases her eggs first, and the male then fertilizes them by covering them with sperm. Spawning requires exact timing of the release of gametes from both sexes, or the sperm will wash away and the eggs will not be fertilized.

Land animals, on the other hand, do not have a watery environment in which fertilization can take place, and so have developed special reproductive fluids to carry the reproductive cells. They also have special organs to effect the transfer. Unlike many aquatic animals, mammals and such land animals as reptiles and birds rely mostly on internal fertilization, which occurs inside the female's body. The males of these animals have special organs that can deposit sperm inside the female during mating. The sperm then swim to the eggs.

Oviparity and viviparity

Although they fertilize internally, most reptiles and birds lay their hard-shelled eggs outside their bodies, and the eggs develop outside the mother. This egg-laying process is known as oviparity. Most higher animals, including most mammals and a few cold-blooded animals, use the form of reproduction called viviparity. In viviparity the fertilized egg implants inside the mother and is nourished by her bloodstream until birth, when it has reached an advanced stage of development and looks like a smaller version of the adult. Some fish, amphibians, and reptiles produce eggs that remain inside the female until the young hatch, but are not nourished directly by her. This process is called ovoviviparity.

Insects

Most insects reproduce sexually and produce yolk-filled eggs, which follow three different patterns of growth before the insect becomes an adult.

The most highly evolved insects, such as butterflies and bees, develop by complete metamorphosis. They hatch from their eggs as larvae that do not resemble their parents. The larvae spend their lives eating and growing. This growth is normally achieved by molting, or the shedding of skin. When the insect is full-grown, it stops eating, and the bloated larva becomes a pupa—such as the chrysalis of a butterfly. The adult insect develops inside the pupal case.

Another pattern of development, incomplete metamorphosis, is exhibited by the group of insects that includes dragonflies and cockroaches. The insects hatch from their eggs as nymphs, or naiads, which resemble the parents, but are wingless, smaller, and sexually immature. Like larvae, nymphs eat all the time. They do not enter a pupal stage, however, but grow and shed their hard outer skin until they reach adult size, when they become sexually mature.

The most primitive kinds of insects, such as the wingless bristletails, hatch as miniature replicas of their parents and no metamorphosis is apparent.

Fishes and amphibians

The reproductive behavior of fishes varies considerably between species. A herring, for

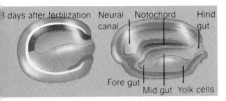

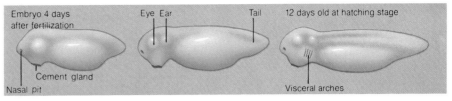

3 days after fertilization Neural canal Notochord Hind gut

Fore gut Mid gut Yolk cells

Embryo 4 days after fertilization Eye Ear Tail 12 days old at hatching stage

Cement gland Nasal pit Visceral arches

12 weeks after hatching

Mouth develops Fore limb

13 weeks after hatching

Develop limbs Tail reabsorbed

Frog

example, may shed tens of thousands of eggs and relies on external fertilization, whereas a dogfish reproduces by internal fertilization and produces very few eggs, which are protected inside an egg purse while they develop.

Amphibians are considered to be more advanced than fishes, but they are not fully adapted to life on land and must always return to water for breeding. Most amphibians fertilize externally. Frogs and toads, for example, mate after they have exchanged specific courtship signals. During the amphibian mating process, called amplexus, the male seizes the female and sits on her. This act induces both the male and the female to release gametes into the water simultaneously. The fertilized eggs hatch as tadpoles, going through a larval stage and gradually metamorphosing into adult amphibians. In some salamanders, however, the young are born live and do not go through a larval stage.

Birds

All birds reproduce by internal fertilization. Unlike amphibians, they do not rely directly on water for breeding. Instead, the hard-shelled eggs they lay each contain a "private pond" in which the embryo floats and develops. The temperature inside the egg is maintained by incubation, in which one or both of the parents, but usually the female, keeps the egg warm. Incubation can take two to three weeks or more before the young chicks hatch. In the case of some birds, the female continues to care for the young in the nest for long periods because the chicks are born blind, helpless, unable to feed themselves, and without feathers to keep them warm.

Mammals

Most mammals fertilize internally and the young develop inside the mother's uterus. However, a few species of mammals, called monotremes, are egg-layers and have a uterus structure similar to that of reptiles. In other mammals, the structure of the uterus varies greatly—female rabbits and kangaroos, for example, have a double uterus and vagina, cows have a two-horned (bicornuate) uterus, and human females have a single, triangular uterus. While the embryo grows inside the mother's uterus, it is supplied with oxygen and nourishment from the mother's bloodstream via the placenta and is kept at a constant warm temperature. After birth, the young mammal is entirely dependent on the mother for milk until it can eat solid foods.

A developing frog embryo (top right and bottom) elongates as it nears hatching, and various sense plates become visible, as do visceral arches and a cement gland in the mouth area. About four days after the tadpole emerges, tail muscles develop and external gills appear around the visceral arches. It attaches itself to plant matter with its oral sucker. About a week after hatching, an operculum grows across the gills, and they wither. The eyes develop and the cement gland disappears. About six weeks after hatching, hind legs bud, followed by forelegs. The mouth then widens and the tail shortens. The metamorphosis is complete at about 13 weeks.

A two-headed Pacific gopher snake is the result of the same reproductive process that creates Siamese twins. It occurs when a fertilized egg splits partially into two during an early stage of the egg's development. Mutations such as these rarely survive because their internal organs are often greatly deformed; they are therefore weaker and more vulnerable than normal members of their species.

Genetics

Genetics is the study of heredity, the means by which offspring inherit physical traits from their parents. Although most offspring produced by sexual reproduction resemble their parents, there is always some variation. Heredity is a complex process in which many different characteristics are transmitted to the offspring, of which only certain ones will show themselves in their physical appearance.

Chromosomes, genes, and alleles

Genes are the carriers of hereditary information. Every individual contains millions of genes, each of which controls one or more characteristics in that individual. Each gene is is made up of a pair of alternate forms, called alleles, and each pair of alleles controls a single characteristic in a normal body cell.

Genes lie on the chromosomes, which are located in the nucleus of a cell. The chromosomes occur in homologous, or matched, pairs—one originally from the mother and one from the father. A sperm or egg cell only carries one of each pair of alleles. This is because during meiosis (cell division) in sex cells, homologous pairs of chromosomes are separated so that each new cell contains only one of a pair of chromosomes and, therefore, one of a pair of alleles. When sperm and egg unite, the new individual inherits one allele for each gene from each parent. Genes on the same chromosome form a linkage group and are usually transmitted together to the offspring. Chromosomes vary in size and shape, their dimensions corresponding to the number of genes—the larger the chromosome, the more genes it contains.

When a gene's two alleles are identical, the gene is said to be homozygous; when they are not, the gene is heterozygous. A heterozygous gene contains one dominant and one recessive allele. Regardless of whether the alleles are homozygous or heterozygous, the dominant gene of a pair of alleles always expresses itself. Recessive genes express themselves only when they are homozygous.

For example, in human beings, the allele for brown eyes (B) is dominant and the allele for blue eyes (b) is recessive. Say, for example, a man who carries a homozygous gene for brown eyes (BB) has children with a woman who has blue eyes (bb). All the father's sperm cells would carry the B allele, and all the mother's egg cells would carry the b allele. As a result, all their offspring would carry the Bb eye-color gene, and because B is dominant, all would have brown eyes. The offspring of two parents with the Bb eye-color gene could either be BB, Bb, bB (all brown-eyed), or bb (blue-eyed). So brown eyes can be produced by either the BB or Bb genotypes. In such simple cases of inheritance, the likelihood of a dominant characteristic occurring is 3:1.

But heredity is not always so simple. Inheritance is not simply a matter of sorting out traits, in which the dominant trait always expresses itself. If this were so, there would only be blue-eyed and brown-eyed people. Instead, many physical characteristics are determined by a number of genes. As a result, characteristics of both parents can mix to produce in the offspring an intermediate result. For example, a parent with blue eyes and a parent with brown eyes may produce a child with green eyes.

When heredity involves the transmission of a pair of characteristics, each of a pair of alleles combines with either member of another pair. This is known as a dihybrid cross. For example, in the fruit fly *Drosophila melanogaster,* if the long wing (WW) and the short wing (ww) characteristics are paired with a broad abdomen (BB) and a narrow one (bb), with one parent being WWBB and the other wwbb, and if 16 offspring were produced, only 1 in 16 would be the same genotype as the parent with recessive genes. But 9 in 16 would bear the dominant phenotypic characteristics—long wings and a wide abdomen, 3 in 16 would carry long wings and a narrow abdomen, and 3 in 16 would have short wings and a wide abdomen.

The chromosomes that determine an indi-

Chromosomes, such as these of a rat-mouse hybrid, show up clearly when stained with a fluorescent dye and then lit by ultraviolet light. In the resting phase, chromosomes are loosely coiled threads of chromatin, but during cell division they spiral tightly, when they become visible, particularly during metaphase. Each species has a definite number of chromosomes—a mouse has 40 and a human has 46.

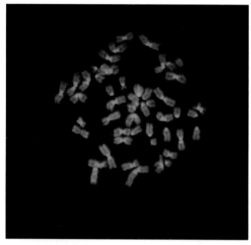

Single-factor inheritance can be seen in the inheritance of coat color in rats. When a black rat with the dominant genotype BB mates with a brown rat of the recessive genotype bb, the resulting offspring will all be genotype Bb and have black coats. But when rats of genotype Bb mate, three out of four of their offspring will have black coats, and only one will have a brown coat. Of the black rats, only one will be homozygous for a black coat (BB), as will the brown rat for its coat color (bb). The rest will be heterozygous (Bb).

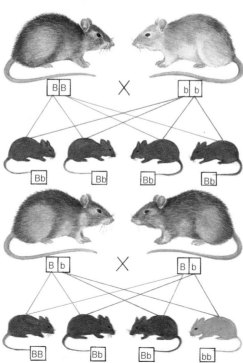

Albinism is a hereditary disorder that manifests itself in an absence of pigments in the body. It is carried by recessive alleles and therefore occurs infrequently. The disorder is not lethal, although it makes life more difficult for victims. For example, lack of protective pigment in the eyes makes them extremely sensitive to light and lack of pigment in the skin increases the danger of sunburn.

vidual's sex are medium-sized. In humans and the fruit fly, the female sex chromosomes are known as X chromosomes and are given the genotype XX. Of a pair of male chromosomes, one is rod-shaped, and the other—the Y chromosome—crooked; this pair is given the genotype XY. In birds, the female chromosomes are XY, and the males XX. In some insects, the female chromosomes are XX, but the males have no Y chromosomes, so their genotype is XO.

During a special stage in meiosis called chiasmata, the alleles leave their chromosome and change places with the corresponding chromosome of the homologous pair. This process results in new combinations of characteristics, known as recombinants, which increase the genetic variation found among individuals.

Some diseases and disorders, such as albinism, in which the skin, hair, and eyes have no pigment, are hereditary. Albinism is not fatal, but some hereditary diseases, such as cystic fibrosis, are. This disease, carried by homozygous recessive alleles, causes the glands to secrete excessive amounts of abnormally thick mucus, and death as a result of intestinal or lung problems usually occurs before adulthood.

The structure of a gene

Genes are made up of a chemical called deoxyribonucleic acid (DNA). The DNA molecule is a double helix, resembling a twisted ladder. Its long parallel strands are constructed from

alternating sugar and phosphate molecules, with cross-bridges formed from special groupings called bases. There are four types of bases: adenine, cytosine, guanine, and thymine, represented by the initials A, C, G, and T. The bases are arranged in specific pairs: adenine always pairs with thymine, and guanine with cytosine. Hereditary information is coded in the sequence of these bases along the DNA molecule. The code is an arrangement of 20 amino acids, the basic components of proteins, which form the structure of living organisms and all the enzymes that control their metabolism.

The white peppered moth is the dominant form of the species. But in areas where tree trunks were blackened by pollution, white moths became easy prey for birds and the black, recessive form flourished. The recent success in cleaning up polluted areas has meant the return to dominance of the white moth.

Behavior

In order to survive, an animal must be able to respond to changing conditions in its environment. In other words, it must adapt. Animals' responses include simple reflexes based on instinct and learning, as well as more complex patterned behavior, such as the division of labor among communities of some insects and the hierarchical system of dominance of such mammal groups as wolves. Mating and feeding are the most important activities, and behavior rituals related to these activities often develop.

Instinct and learning

All animals are born with a range of inborn behavior patterns. The knowledge of how to do certain things, called instinct, is passed from parent to offspring genetically. This form of inherited behavior helps an animal survive in its environment.

In addition to instinct, animals are able to learn ways of coping with their environment and can thus modify their behavior to deal with problems that they have encountered before. Some scientists who study animal behavior consider this ability to learn to be a measure of an animal's intelligence.

Most animals use a combination of instinct and learning to adapt to their environment; the longer an animal's life span and the more complex its life style, the larger role learning plays in its life. A large predatory mammal, for example, has a built-in hunting ability that is improved by learning. A fly, by comparison, which lives for only a short time, needs only to feed and breed successfully before it dies. It does not have the time to learn how to find a mate, or where it would be best to lay its eggs. Instinct provides the fly with a fixed set of adaptive responses so that it can cope even though it has no previous knowledge of a situation. Instinct is also important when there is no parental care, when the animal is not able to learn by example.

Instinctive reaction is characterized by two features: it is the same for all members of a species, and it can be initiated by a simple stimulus. But the division between instinct and

A rat placed in a Skinner box (A) will quickly learn how to get what it needs from its surroundings. It learns that if it waits (B) for the stimulus—the flashing light—then presses the lever (C), food drops out of the chute, and it can eat (D).

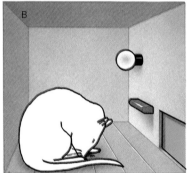
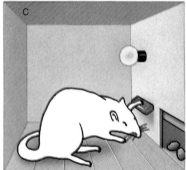

The courtship ritual of gannets, which breed in large communities, is essential to the male in the first breeding year. Because male and female gannets look alike, behavior enables a male gannet to determine the sex of any apparently unmated bird. In the ceremony two birds stretch and twist their necks. In subsequent breeding seasons the partners will recognize each other and come together again.

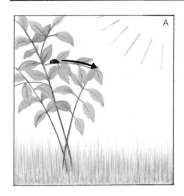

Ticks are drawn toward sunlight (A) by the smell of butyric acid, given off by warm-blooded animals (B), and the warmth of flesh into which they sink their proboscis and suck up blood (C). Scientists have found that a tick will try to suck a warm, water-filled balloon onto which butyric acid is smeared.

learning is not immediately obvious. All animals of a particular species show the same reaction to a particular stimulus, even if they have been isolated from birth, although animals also often carry out learned activities in a similar stereotyped fashion, which may appear to be instinctive. Responses learned in isolation vary, however, among individuals of a species, depending on how they were taught.

An animal's ability to learn and remember is, to a certain extent, dependent upon the life it leads. A worker bee, for example, must be able to learn the position of its hive and the location of flowers that are good for foraging, but a fly does not need this ability. It is not surprising then that experiments have proved that whereas bees can easily remember the position of plates of food, flies are less able to do so.

Insight is probably the highest form of learning—when the answer to a problem arrives in a sudden flash. The term is used to describe the rapid solution of a problem, skipping the process of trial and error. An example of this is the immediate construction by some chimpanzees of a "ladder" from objects in order to retrieve fruit that is too high to get at by stretching their arms.

Another type of learning is imprinting, which is thought to occur only over a short, critical period when the animal is very young, or at crucial times of an animal's life, such as at the production of offspring. Imprinting occurs in both mammals and birds. Young songbirds, for example, are not able to sing an adult version of the species song if they have never heard it. They will sing, instead, a simple version of the song, without any characteristic trills. But a young bird that has heard the adult song only once will be able to sing it perfectly at a later stage because it has imprinted the song.

Small animals and birds also imprint any object that is near them soon after birth. This "maternal" imprinting ensures that the young follow the mother, as waterfowl do. But young birds can be caused to imprint balloons, puppets, cardboard boxes, and even human beings.

Apart from maternal imprinting, there is also sexual imprinting. An animal that has, for some reason, managed to imprint the wrong species, or an inanimate object, will not mate with its own kind. This imprinting was illustrated in an experiment in which a male zebra finch was raised by a pair of Bengalese finches. When the zebra finch reached maturity, it was placed in a cage with a female zebra finch and a female Bengalese finch; the

male ignored the female of his own species and courted the Bengalese female, despite her lack of interest.

Scientists believe imprinting is largely based on smell and may account for a mother's recognition of her offspring. For example, when cows and wildebeests give birth, the mother imprints the smell of her offspring and is then able to identify it in a herd.

The role of sense organs

In addition to instinct, learning is closely associated with responses that result when sense organs are stimulated. Most animals are aware of and respond to changes in light, gravity, temperature, and noise, but they do so in different ways.

Many animals forage for food, then bring it back to their home to eat. However, they need to orient themselves in their environment and recognize signs to find their way back. Digger wasps, for example, recognize the pattern of landmarks around their burrows. If the pattern were changed, they would not immediately recognize the position of their burrows.

Stimuli such as those used in courtship displays are often spread over a few hours or days and have a cumulative effect on an animal. These stimuli break down defensive barriers between the male and female and also bring the female into a sexually receptive state

A feeding animal will generally allow competitors to approach only within a certain distance—the boundary of its "feeding territory." Here a lioness has even temporarily abandoned her kill to rush at scavenging hyenas and frighten the intruders away. Only when she has eaten her fill will they be allowed to pick over the remains.

Combat among the males of a herd is often the means of asserting rights over territory, females, and the dominant position in the herd. Among animals such as Thomson's gazelle, males and females form separate herds. A few dominant bucks lead the bachelor herds, and during the breeding season they try to attract females into their territory so that they can mate with them.

so that mating can take place. The breakdown of defensive behavior is necessary in certain species that instinctively attack or run from any other living thing. If a male scorpion or spider intending to mate does not approach the female cautiously, for example, she may attack or kill him.

Animals respond to stimuli that are quite different from those that humans can perceive. For instance, for many years zoologists wondered why some flowers that depended on bees for pollination were completely white and apparently had no guiding pattern (nectarguides) that would lead the insects to the nectar source. The scientists later discovered that bees can see ultraviolet light, and that some flowers have very distinctive ultraviolet patterns.

Color also plays a part in stimulating sexual behavior. For example, during the mating season, the belly of the male stickleback fish becomes red and his back turns blue-white. This coloration attracts females, which during the mating season are plump with eggs. The male guides a female to a nest that he has constructed. She enters it, and he nudges her from behind, causing her to release the eggs. She then swims away, and the male enters the nest and fertilizes the eggs.

Many visual stimuli to which animals respond are crude. For example, a robin will attack a clump of red feathers in its territory during the breeding season, and some cuckoos automatically feed a gaping mouth whether it is that of a fish or a baby bird. But sometimes animals respond to an artificial stimulus more vigorously than a natural stimulus. For example, if offered an egg considerably larger than their own, oyster catchers and other birds will try to hatch it instead of their own egg. Herring gull chicks peck at a red patch on the lower part of the parent's yellow bill, which causes the parent to regurgitate food. But scientists have found that a red and white striped pencil causes the chicks to peck more vigorously than a real gull bill. This indicates that the chicks prefer to peck at a bill with a high level of contrast. This is a case of misfiring, when the animal's behavior is stimulated under inappropriate circumstances and the animal's needs are not met.

Scent organs are essential for most animals to locate scent trails to food sources, but scent is also important in communicating fertility. At the time of estrus, the period when a female is able to become pregnant, many female mammals secrete highly scented body chemicals called pheromones, which the males can detect. The phermones emitted by some insects can be detected at a distance of several miles.

Parental behavior

Animals have many different ways of rearing young. In some ducks, for example, the male deserts the female to mate with others in that season. In sticklebacks and other types of fish, however, the female leaves after spawning, and the male protects the eggs and cares for the young until they become independent. But among many birds and some mammals both parents care for the young, which may be unable to feed themselves or maintain an adequate body temperature for several weeks after birth.

Ties between parents and their offspring are not necessarily exclusive. As many as a hundred pairs of sociable weavers raise their chicks together in a large nest, and many primate groups work together to protect the young from predators. Prairie dogs go a step further. A young prairie dog may be suckled by any lactating (milk-producing) female in the community and groomed by any male.

Feeding

The life styles of all animals are organized chiefly around their method of obtaining food. Some parasites (animals that live in or on the bodies of other animals), for example, modify the behavior of their host to ensure that they will go on to the next stage of their life cycle. For example, some fish parasites cause their hosts to swim closer to the surface of the water than those free from parasites. This makes it more likely that the fish will be eaten by birds, which then become the hosts for the next stage of the parasites' life cycle.

Many animals feed in groups. This behavior provides the animals security because it is more likely that a predator will be detected by one member of the group sooner than if the animals are feeding separately. In addition, animals grazing in a group look up less often than when they are on their own, which means that each individual can spend more time eating.

Social behavior

Some animals are solitary all the time, and others are solitary most of the time but come together for certain activities, such as migration. A species of locust has a solitary phase, when its coloring is green. But when environmental conditions are favorable, the insects change color to black and russet and become part of a migratory swarm. Some birds, such as the Eurasian robin, will attack any member of their own species that comes near their territory, but will join a flock that migrates as winter approaches.

Other animals gather and coordinate their activities. Termites, ants, and bees, for example, live in highly structured societies, or colonies. The division of labor between the different animals in the colony is strict, and the number of different kinds of individuals depends on what the colony requires for its survival. The colonies are controlled by pheromones that are produced by the queen. These chemicals and the high degree of physical contact between members of a hive ensure that the community works together effectively.

Most primates live in groups. These groups are less regimented than those of social insects, but order is maintained by a social hierarchy, which is headed by a single dominant male, usually the oldest and largest. This male has the first choice of the best food and most receptive females. Aggressive confrontations are usually avoided by the use of a large number of facial expressions and dominant or submissive gestures.

Animals that live in close communities often hunt together. Lions, for example, hunt by stalking and then ambushing their prey. But hunting dogs run down their prey in packs.

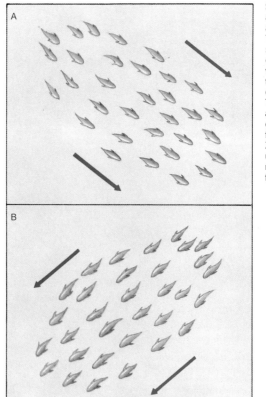

There is usually no dominant dog, male or female, although one dog will determine the direction of movement of the group. The dogs return to the lair after the kill, where the pups wait, and the whole pack feeds on regurgitated meat.

Animals such as hunting dogs, which have a high-protein diet, can afford to spend a lot of time in activities such as play because the energy value of the food they consume is high. Play involves romping and mock fighting and is important to young animals because through play they learn skills that they will need as adults.

Schooling fish are remarkable in the way they travel and communicate in shoals. The shoals are elliptical in shape (A), and contain fish of roughly the same size, which travel at equal distances away from one another, at the same speed. There does not seem to be a leader, and a change of direction is made by all the fish at the same time (B). Zoologists do not fully understand how this communication between the fish occurs.

Grooming is practiced by most mammals, but among primates, particularly Old World monkeys, it reflects rank order and serves to maintain the hierarchy of the group. Subordinate individuals spend more time grooming dominant clan members than being groomed themselves. Male baboons, for example, lie in front of females, which are expected to groom them. The males groom back, but for not as long. Grooming is also an indication of reassurance and friendliness.

habitats of the world

The desert supports a variety of animal life highly adapted for a hot, dry environment. The gila monster, a large poisonous lizard of North America, stores fat in its thick tail and can live on this fat for months when food is scarce.

Zoologists study animals in many different ways. One useful way of thinking about animals is according to where they live. Of course, where an animal lives—its habitat—is related to its anatomy. Physical traits that enable an animal to survive in a particular place may be passed to offspring. Natural selection may enable more of these offspring to survive than do offspring that lack these traits. Such helpful adaptations may spread to the whole species over time.

Seven major habitats are found all over the world. Each habitat supports many kinds of animals.

Tropical rain forests

These warm, humid forests are home to more species than any other habitat on Earth. Monkeys, bats, tree boa constrictors, colorful parrots and macaws, and wasps and beetles live in the overhanging layer of foliage called the rain forest canopy. Mammals such as jaguars, tigers, tapirs, and anteaters prowl in the forest below. In African rain forests, chimpanzees and lowland gorillas spend their lives between the ground and the trees. Crocodiles and fish swim in rivers and ponds.

Grasslands

The world's grassy plains areas support herds of herbivores such as gazelles, gnus, zebras, and bison. Carnivorous predators such as cheetahs, lions, and hyenas roam the African grasslands—called savannas—searching for such prey. Other animals of the African savannas range from the elephant, rhinoceros, and hippopotamus to the giraffe, ostrich, and termite. Animals of other grasslands include the Australian kangaroo and wombat, the South American cavies and increasingly rare pampas deer, and the North American coyote, pronghorn, and prairie dog.

Deserts

Some animals are adapted to survive in the harsh, dry climates of the world's deserts. These include centipedes and scorpions; insects, such as bees and butterflies; reptiles, such as rattlesnakes and Gila monsters; birds, such as elf owls and roadrunners; and mammals, such as kangaroo rats, mule deer, and dingoes. Many desert animals have light-colored skin, which reflects sunlight. Some, such as the desert fox and the desert hare, have large ears that help them get rid of body heat.

Temperate forests

Unlike tropical rain forests, temperate forests often have thick underbrush. Mammals small enough to move through this growth include chipmunks, opossums, raccoons, and squirrels. Larger forest mammals include various species of bear and deer. The koala lives in the temperate forests of Australia. Numerous birds of different species nest in the trees and shrubs. Salamanders are common, as are snails. Fish, frogs, and turtles live in forest ponds and streams.

Mountains

The mountains are home to many animals, and the species vary with the altitude. At lower—and therefore warmer—elevations, bear, deer, elk, llamas, vicuñas, and yaks find food and shelter. Spiders and insects, such as butterflies and grasshoppers, also live at this level. Above the tree line, where the climate is colder and the air is thinner, bighorn sheep and mountain goats thrive on the rocky slopes. High on the snowy peaks, only a few insects, spiders, and ice worms can survive.

Polar regions

Animals of the cold, treeless northern plains of the tundra include thick-furred caribou and musk oxen. The short ears of the arctic fox and arctic hare help these animals retain body heat. Mosquitoes and other insects breed in shallow tundra ponds. Only a few species of insects and ice worms can live in areas that are snow-covered year-round, but the cold waters of the Arctic and Antarctic teem with fish, sponges, whales, and tiny shrimplike krill.

Many kinds of animals live in mountains, and the species vary with the elevation. Guanaco, wild wool-bearing animals of South America, roam the dry foothills of the Andes Mountains up to 14,000 feet (4,300 meters).

Oceans

The ocean is home to some of the smallest animals in the world as well as the largest—the blue whale. Some lobsters, sea urchins, and numerous species of tropical fish prefer warm tropical ocean waters while live anemones, barnacles, and starfish are found along the shore. Even in the dark, cold waters of the ocean depths, such specially adapted animals as the anglerfish survive.

Coral reefs are home to a multitude of the ocean's creatures. Barnacles, urchins, sea anemones, and other sea animals grace the rocks and limestone formations of the reef, while colorful fishes dart in and out of the abundant coral and vegetation.

Invertebrates

Of the million or so known animal species, more than 98 per cent have no backbone and are classed as invertebrates. They have a wide variety of shapes and sizes, from tiny animals only a fraction of a millimeter in length to the giant squid, which may measure up to 55 feet (17 meters) long. Invertebrates have an extraordinary range of life styles, and occur at every level of the complex food web that links all forms of life.

Classifying invertebrates

Scientists group invertebrates in a number of ways besides the formal organization into phyla, orders, genera, and species. The first is according to whether the body is bilaterally symmetrical or radially symmetrical. An animal that is bilaterally symmetrical, such as an earthworm, has matching right and left sides and has recognizable front and back ends. An animal that is radially symmetrical, such as a sea star, is wheellike and can be divided into two matching halves by any line that passes through the center. Radially symmetrical animals have no recognizable front or back.

Scientists also distinguish invertebrates according to the number of cell layers they have. A few invertebrate phyla are made up of species whose bodies are composed of only inner and outer cell layers. These are the simplest of all multicellular organisms.

The bodies of most invertebrates, however, consist of three tissue layers. In addition to inner and outer layers, they have a third layer in the middle. These types of invertebrates are further classified according to whether they have a coelom—an internal, fluid-filled body cavity.

The coelom provides a space where organ systems typical of higher animals can develop. Animals with a true coelom are called coelomates. Animals that have a primitive coelom are called pseudocoelomates. Animals that lack a coelom are classified as acoelomates. The area between the outside of the body and the acoelomate's gut is solid tissue.

Finally, scientists describe invertebrates by habitat and life style. Unlike vertbrates, which can all move freely from place to place, some invertebrates are sessile, or fixed in one spot. Others are described according to whether they eat meat or plants, whether they filter food from the water, or whether they extract food from sediment. Many invertebrates, such as certain worms, are parasites, which live in or on the body of a plant or another animal. Others live symbiotically with another organ-

The zebra spider, a jumping spider, is a member of the phylum Arthropoda—the largest group of animals, containing about 900,000 species. Jumping spiders can leap a short distance; they hunt primarily by sight and have the best vision of all spiders. Their eight eyes, arranged in two rows of four, consist of a large number of receptors and can perceive a sharp image of considerable size. In addition to their keen eyesight, the spiders have chemosensitive hairs at the tips of their appendages.

ism. In a symbiotic relationship, both members benefit.

Types of invertebrates

There are about 35 invertebrate phyla, but because new groups are continually being discovered and the classification of invertebrates is constantly under revision, the exact number and distribution of phyla is debatable.

Nine groups are particularly important because they represent about 90 per cent of all living invertebrates. These nine phyla are: Porifera (sponges), Cnidaria (jellyfish and related species), Ctenophora (comb jellies), Platyhelminthes (flatworms), Nematoda (roundworms), Mollusca (snails, squids, and clams), Annelida (earthworms, ragworms, and leeches), Arthropoda (insects, spiders, and decapods), and Echinodermata (sea stars and sea urchins).

The simplest of these phyla are the poriferans. In fact, scientists once considered them as plants because they lacked so many of the characteristics of higher animals. Cnidaria and Ctenophora are slightly more advanced, with a mouth but no specialized tissues for digestion or excretion. Platyhelminthes are slightly more highly developed than the cnidarians, having three layers of body cells. Most platyhelminthes, however, lack a body cavity, a circulatory system, and an excretory system. The annelids, or segmented worms, which include earthworms as well as numerous species of aquatic worms, are the most highly developed worms.

All the animals in the phylum Echinodermata are based on the same radial structure,

Sea slugs are gastropod members of the phylum Mollusca, which also includes snails, limpets, abalones, and slugs. They are among the most brilliantly colored of all marine invertebrates and, unlike most other mollusks, they have no shell. Some have colored tentacles called cerata along their back.

usually consisting of five or ten arms radiating from a single mouth. They live only in the ocean and are all slow-moving or fixed to one spot and unable to move around.

The arthropods are the most successful invertebrates in terms of numbers of species: the arthropod phylum contains more than three-fourths of all the different kinds of animals. All arthropods have jointed limbs and segmented bodies covered by a hard exoskeleton. This body covering acts as a protective armor and as a frame onto which muscles can attach.

The largest class of arthropods is Insecta, the insects, which are primarily land-dwelling. Crustaceans, the second largest class of arthropods, range from the relatively advanced lobsters, crayfish, and crabs to the tiny copepods that graze on the phytoplankton in the oceans. The third group comprises the arachnids—spiders, ticks, scorpions, and related species. Of the six remaining classes of arthropods, only the centipedes and the millipedes are represented in great numbers.

Mollusca is the second largest phylum, after the arthropods, containing more than 100,000 species. They are probably the most highly developed of all invertebrates. Most mollusks have a hard shell enclosing a soft body, and although they conform to a general anatomical pattern, their external body forms are extremely varied.

The minor phyla

The remaining phyla include the smallest phylum, Placozoa—with only a single species—and Priapuloidea, which contains only nine known species of tiny, primitive cucumber-shaped, seabed-dwelling worms. Most of the small phyla comprise marine species that are sand or seabed dwellers.

Some invertebrates seem to form a transition between the invertebrates and the vertebrates. These are the hemichordates, such as acorn worms; the tunicates, including sea squirts; and the cephalochordates, or lancelets, which are members of the phylum Chordata. Animals of the phylum Chordata have a simple skeletal rod, or notochord, which is considered an earlier stage of the spinal column.

Protozoa and sponges

Protozoa, or "first animals," are simple, unicellular (single-celled) organisms that make up the kingdom Protista. Although protists are not technically animals, they have a few animallike characteristics, and many scientists believe that higher animals evolved from them.

Protozoa are divided into four subphyla: the Sarcomastigophora (amebas and flagellates), the Ciliophora (ciliates), the Sporozoa, and Cnidospora. Most protozoa are microscopic.

Scientists have identified more than 100,000 species of protozoa. They are found in almost every habitat where moisture is present and also as parasites in most animals. Because of their resistant spores, some protozoa can withstand extremes of temperature and humidity.

Amebas and flagellates

Amebas have a constantly changing body shape and move by producing pseudopodia (false feet). The cytoplasm of the ameboid cell is pushed out to form the pseudopodium as the animal moves forward. The common *Ameba* has a naked cell surface, but a variety of shelled forms exists. The genus *Difflugia*, for example, constructs a case from grains of sand, whereas other amebas secrete shells of calcium carbonate and silica. These shells contain holes through which the pseudopodium can be extruded to collect food particles.

Some amebas are parasitic, such as *Entamoeba coli*, which lives in the human intestine. Found in up to 30 per cent of the world population, it scavenges bacteria and food particles. This harmless association is termed commensalism. But *Entamoeba histolytica* is harmful and causes amebic dysentery.

In contrast to amebas, flagellates move using a long, hairlike structure called a flagellum, which beats like a whip to provide propulsion. Most flagellates have a fixed body shape (usually oval), and almost all reproduce asexually by binary fission; but there is considerable diversity between species. Choanoflagellates, for example, have a delicate collarlike structure that surrounds the base of the flagellum and helps to collect food particles. Similar collar cells (choanocytes) are found in sponges, which suggests a possible link between these two groups. In some collared flagellates, the individuals do not separate after cell division, but give rise to a colony.

Most flagellates feed on small particles and organic materials dissolved in the surrounding water. But some, such as the green *Euglena* commonly found in ponds, are able to produce their own food by photosynthesis. The presence of the photosynthetic pigment, chlorophyll, has led many biologists to classify

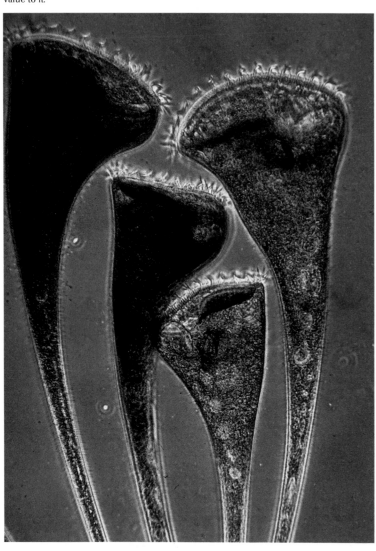

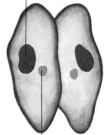

Macronucleus

Micronucleus

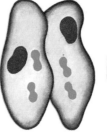
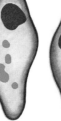
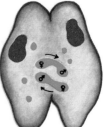

Synkaryon

The ciliate *Paramecium* reproduces by conjugation. When two adults with compatible micronuclei come together, the micronucleus of each divides twice. Three of the four new cells disintegrate but the fourth continues to divide. The cell walls joining the two ciliates dissolve, and the male nuclei of the two conjugants are exchanged. The adults then separate, and development of the new cell, called the synkaron, continues in each one. At about this stage the macronucleus starts to disintegrate and is resorbed. The synkaryon di-

such flagellates as algae (simple plantlike organisms belonging to the kingdom Monera). In all other respects, however, they are identical to flagellate protozoa.

Many flagellates live in association with other animals. *Trichonympha,* for example, lives in the intestines of termites. The termite relies on the protozoan to digest the wood that it eats—a mutually beneficial relationship known as symbiosis. But some flagellates are blood parasites and cause disease. *Trypanosoma,* for example, causes African sleeping sickness.

Ciliates

The most complex and diverse species of protozoa owe their name to the cilia (short hairlike fibers similar in structure to flagella), which grow in orderly rows on the body and beat rhythmically to propel the animal. Ciliates reproduce asexually by binary fission, or sexually by conjugation. They differ from other protozoa in that they have two nuclei—a macronucleus and a micronucleus—instead of one.

Sporoza and Cnidospora

These organisms, the spore-formers, have no distinct adaptations for movement because they are all parasitic. They live in all animals and are often transmitted by insects. Their name comes from the production of spores, or cysts, during the infective stages of their life. The life cycle of this group is complicated—reproduction alternates from asexual to sexual. Asexual reproduction involves the binary fission of spores, usually in the host. The offspring develop into gametes, which mature and fertilize other gametes that eventually produce spores.

Sponges

The simplest group of invertebrates is Porifera (porebearers), the sponges. Scientists estimate that the number of poriferan species ranges from 5,000 to 10,000 species and that 98 per cent of these live in salt water. Most sponges are found in shallow waters, although some live in deep water. Sponges range in length from a quarter of an inch (0.5 centimeter) to more than 4 feet (1.2 meters). They lack muscles and organ systems of any kind and spend their lives cemented to a solid surface, or substrate, filtering tiny food particles from the water.

Sponges consist of several cell types, each of which performs a specific function and is independent. This feature means that regeneration is easy—if a sponge is fragmented, the cells simply organize themselves into a new sponge.

The simplest sponges are tubular with an external layer of epithelial, or lining, cells. The internal surface is covered with collar cells, or choanocytes, equipped with whiplike structures called flagella, which maintain a flow of water through the sponge. Sponges extract food particles from this current, which they also use for gas exchange and waste removal. The sponge draws water in through small pore cells in its walls, and ejects it from its large mouth, or osculum. More advanced sponges have complex systems of canals and chambers through which water is channeled.

Sponges reproduce both sexually and asexually. Most are hemaphroditic, or have both male and female characteristics, and produce eggs and sperm at different times. Asexual reproduction occurs by budding. In addition, some sponges produce clusters of cells surrounded by a protective covering called gemmules that survive when the parent body disintegrates in winter. In spring the gemmule develops into an adult sponge. Sexual reproduction occurs when sperm is released and enters another sponge in the water current.

Glass sponges, or hexactinellids, comprise the class Hexactinellida, and are mainly deepwater sponges. Like other sponges, they have special cells that secrete the skeleton, which is made up of spicules, or tiny needlelike structures. In bath sponges, these spicules are composed of a horny protein called spongin, but in glass sponges they are made of silica. These siliceous spicules are fused to form a six-pointed shape, from which the class name of the glass sponges is derived.

vides three times, producing eight new cells. Three of these dissolve, four start to attain macronucleus status, and one continues to divide

as a micronucleus. The ciliate itself then starts to divide. The four macronuclei polarize, two at each end, and when the cell finally di-

vides it also contains one micronucleus. The micronucleus in both new ciliates continues to split and the macronuclei again polarize,

so that when the two ciliates divide, the resulting four cells each contain one micronucleus and one macronucleus.

Cnidarians and ctenophores

The cnidarians and ctenophores are among the simplest of the multicellular organisms

The sea nettle, common in the coastal waters of the Atlantic Ocean and Mediterranean Sea, has a diameter of 12 inches (30 centimeters). Its fringing tentacles and its four trailing arms are covered with nematocysts, or stinging cells, with which sea nettles can deliver a painful sting to swimmers.

In the reproductive cycle of most jellyfish, the egg released by the adult medusa becomes a sausage-shaped larva covered with cilia called a planula larva. The planula larva develops into an attached polyp. The planula then settles on a surface and turns into a polyp form called a scyphistoma. As the scyphistoma develops, its body begins to divide into flattened sections. At this stage, it is called a strobila. Each section breaks free and swims off as a separate individual, called an ephyra, that eventually becomes a mature jellyfish.

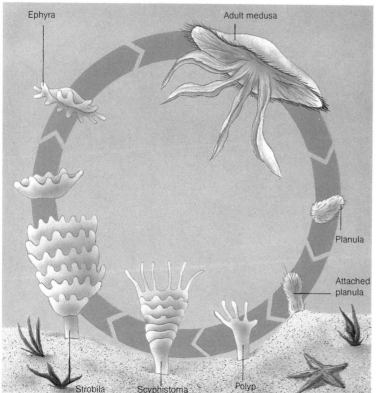

Ephyra

Adult medusa

Planula

Attached planula

Strobila   Scyphistoma   Polyp

that have cells organized into tissues. Cnidarians comprise the phylum Cnidaria and include such animals as corals, jellyfish, hydras, and sea anemones. The largest species is the arctic jellyfish—7 feet (2.1 meters) across with tentacles 100 feet (30 meters) long—and the smallest are the individual polyps of some coral colonies, most of which are less than one inch (2.5 centimeters) in diameter. Most cnidarians are marine animals, although there are a few freshwater species. Scientists have identified about 10,000 species of cnidarians, and they are probably the most common macroscopic marine animals, especially in tropical and subtropical coastal waters.

Scientists have described about 100 species of ctenophores, which make up the phylum Ctenophora. All of them live in the sea and most drift about with the ocean current.

General structure of cnidarians

The cnidarians have two layers of cells that surround a tubular body cavity, with an opening at one end forming a mouth, and no specialized tissues for breathing or elimination of wastes.

Cnidarians have two basic forms—polyps and medusas—which are radially symmetrical, with no definite front or rear ends. The flower-like polyps are attached at their base to their mother organism, and have a mouth on the upper side, surrounded by tentacles. Medusas, commonly known as jellyfish, are free-swimming, bell-shaped organisms, with a mouth on the underside. In the typical life cycle of a cnidarian, the two basic forms alternate—polyps bud off medusas, which then produce a larval polyp, and so on. One form is usually dominant, however, and in some species the other is omitted altogether.

The cnidarian body is made up of two layers of cells—an outer epidermis and an inner gastrodermis—separated by a layer of jellylike matter called the mesogloea. The two layers of cells surround a central digestive cavity, or coelenteron, and contain muscle cells that contract to bend the body or draw in the tentacles. In cnidarians the epidermis contains stinging cells called nematocysts to immobilize prey. The tentacles then push the food through the mouth into the coelenteron, where it is digested by enzymes and absorbed by the gastrodermis.

Many cnidarians form colonies of thousands of individuals. In the simplest colonies all individuals are identical, but feed and reproduce separately. In more advanced colonies, however, such as those of some corals and hydrozoans, several different forms are present, and each type undertakes a different function.

Hydrozoans

The class Hydrozoa has about 3,000, most of which live in salt water. Some hydrozoans, such as the common *Hydra,* live in fresh water. In most hydrozoans, symbiotic algae called zoochlorellae live in the gastrodermal cells of the body walls and give hydrozoans their characteristic green coloring.

Hydrozoans generally live as both medusas and polyps. Hydra and a few other species, however, have no medusa stage, and some hydrozoans display the medusoid form only. In the reproductive cycle of most hydrozoans, sexually reproductive medusoids bud off the parent and either drift free or remain attached to the parent organism, in which case they are known as gonophores. *Hydra,* however, more commonly reproduce by budding from the parent.

Most hydrozoans live in colonies of many individuals. Although they normally live attached to a surface, they can move by a creeping action at the base. Many hydrozoan colonies, such as the Portuguese man-of-war are attached to an air-filled cushion that acts as a float. Its tentacles are made up of a colony of polyps that is polymorphic, or made up of two or more types of individuals.

Jellyfish

Jellyfish (class Scyphozoa) consist of a swimming bell fringed by tentacles, with a four-cornered mouth on the underside. Around the mouth are four trailing arms, which, like the tentacles, are well supplied with nematocysts. Jellyfish swim by contracting and releasing a ring of muscle cells. Balancing organs called statocysts and simple light receptors around the edge of the bell help jellyfish remain upright. As their common name suggests, they contain large amounts of jellylike mesogloea, which helps to control buoyancy and also acts as an elastic support for the body.

Sea anemones and corals

The sea anemones and corals of the class Anthozoa (flower animals) have a polyp stage only. These polyps are often large and complex, with the coelenteron divided by partitions, or septa. Many sea anemones have large attachment disks and thick, leathery bodies, which allow them to survive on rocks that are exposed to the air at low tide.

Corals are essentially polymorphic colonial sea anemones, although solitary forms exist. The polyps secrete skeletons, the exact form of which defines the species, such as the delicate sea fan. Like hydrozoans, corals have photosynthetic algae called zooxanthellae living in their gastrodermal cells. Because of the algae's need for light, reef-building corals flourish only in tropical and subtropical coastal waters where the water is less than 100 feet (30 meters) deep. These corals also require temperatures above 65° F.(18° C).

Ctenophores

Ctenophores, or comb bearers, share several characteristics with cnidarians. They are radially symmetrical, jellylike, and composed of two layers of cells. But the cnidarians' medusoid shape has been adopted and modified into a sphere or oval. Ctenophores get their name from their comb plates, which form from fused cilia. There are two classes of ctenophores: those with tentacles, such as the most common form, the sea gooseberry, and those without. These animals are usually found drifting among plankton, and many are luminescent. They swim through the water by beating their comb plates consecutively from the head to the tail. Ctenophores catch their prey using sticky "lasso cells," called colloblasts. Colloblasts have a similar function to that of nematocysts in cnidarians.

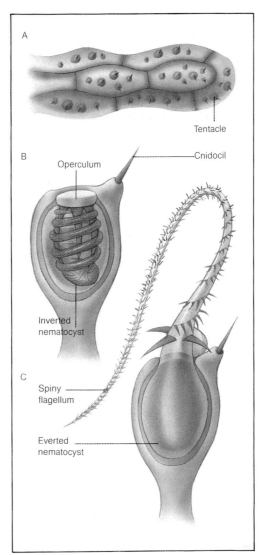

A

Tentacle

B

Operculum

Cnidocil

Inverted nematocyst

C

Spiny flagellum

Everted nematocyst

Nematocysts (stinging cells) are abundant on the tentacles of cnidarians (A). These oval cells contain an inverted tube, at one end of which is a flagellum, often covered with spines, coiled tightly around the bulb of the tube (B). When the protruding hairlike trigger (cnidocil) is activated, the cell lid (operculum) snaps open, the nematocyst whips out (C), and the toxins contained in it are ejected through the spines.

Sea anemones, like most cnidarians, are carnivorous, the larger ones feeding on small crustaceans. The nematocysts in their tentacles do not have a cnidocil, or an operculum; they often simply break through the capsule wall. The tentacles surround an oral disk that contains a mouth. Prey is carried to the mouth, which is widened by muscles, and the prey is swallowed whole.

Platyhelminthes

The phylum Platyhelminthes (flatworms) is divided into four classes: the Turbellaria, or non-parasitic flatworms; the Monogenea, or parasitic flatworms; the Trematoda, or flukes; and the Cestoda, or tapeworms. As their common name suggests, platyhelminthes are flattened, soft-bodied organisms. Most turbellarians are free-living. The members of the other three classes are exclusively parasitic. Most flukes are microscopic. Intestinal tapeworms, on the other hand, may grow up to 100 feet (30 meters) long.

General features

Compared with the simpler cnidarians, flatworms show several advanced characteristics. For example, the body is bilaterally symmetrical and has a definite head end. Free-living flatworms are active animals, and in many species the head carries pairs of eyes, as well as organs that sense chemicals. There is also a concentration of nerve cells at the front end of the body that forms a primitive brain, in addition to a nerve net similar to that of the cnidarians.

The flatworm body is composed of three layers of cells. A layer of mesodermal cells lies between the epidermis and the gastrodermis, which lines the digestive cavity. Because these animals are so highly flattened, a specialized breathing system is not required and oxygen reaches this middle cellular layer by diffusion. The mesoderm contains the complex reproductive organs, which are made up of different types of cells. This differentiation represents a higher level of organization than that found in cnidarians and similar phyla, which possess tissues, but not organs.

As in hydra, the digestive cavity in flatworms (other than tapeworms) has one entrance only and no separate exit, with the result that undigested food is ejected through the mouth. In turbellarians, this digestive cavity is often highly branched, so that food can be distributed to all parts of the body.

Turbellarians

The turbellarians include all free-living flatworms. Some species grow up to 25 inches (60 centimeters) long, but most are about one-half inch (13 millimeters) long. Most freshwater species are drab and inconspicuous, whereas tropical marine species may be very colorful.

Although they live in water, turbellarians do not swim freely, but creep along the bottom. The epidermis is equipped with tiny, hairlike cilia and glandular cells, which secrete mucus in which the cilia beat, enabling the turbellarian to glide along. Turbellarians also have bands of muscles that allow complex bending movements of the body.

Most turbellarians are carnivorous, feeding on small animals, or necrophagous, feeding on dead animals. Many secrete mucus and an adhesive to entangle their prey, then break it up into small particles using a part of their gut, which they can extend out of their mouth. In some cases, the prey is ingested whole.

Compared with parasitic flatworms, turbellarians have a simple life cycle. Most species are hermaphroditic, with male and female sex organs in the same individual. Self-fertilization does not usually occur, however, and a partner is needed for fertilization to take place. Sperm is exchanged between the two individuals, and both lay fertilized eggs in egg capsules or gelatinous masses. After two or three weeks, young flatworms resembling their parents hatch from the eggs.

Some turbellarians also reproduce asexually; the body constricts in the middle and the

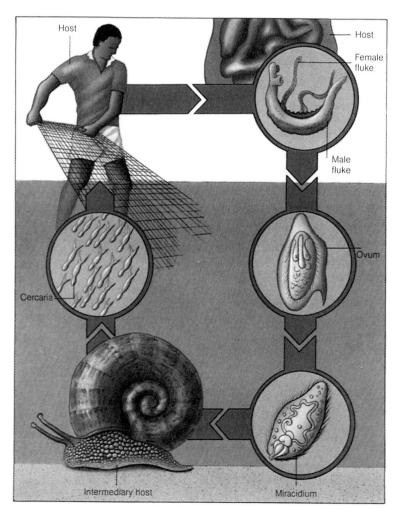

The blood fluke breeds in the host's intestine or bladder. Fertilized eggs pass out in the feces, develop into larvae (miracidia), and are released in water. In an intermediary host they breed cercariae, which burrow into a third host.

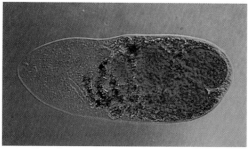

The small intestinal fluke (above) is one of many that are found particularly in Asia. This species has only one intermediary host, usually snails or fishes. It is transmitted to vertebrate hosts that eat raw fish and mollusks.

hind portion breaks off to form a complete new individual. Using the same principle, if an individual is cut in two, the front half will form a new tail and the rear half a new head.

Flukes

These flatworms occur as parasites that live inside other organisms (order Digenea) and parasites that live outside their hosts (class Monogenea). Flukes generally have leaf-shaped bodies with suckers or, rarely, hooks to attach to their hosts. As a result of their parasitic habits, adult flukes lack the sense organs and layer of cilia that are found in turbellarians.

Another feature found in flukes and typical of most parasites is a high reproductive ability, which ensures that at least some offspring reach suitable hosts. Not only do the adult flukes of the order Digenea produce large numbers of eggs, but the larvae of these flukes also reproduce within their hosts.

Monogenea are often found in the gill cavities of fish, where they have to withstand strong water currents. They feed on blood and gill tissue. Internally parasitic flukes may be found in many vertebrates as well as in other invertebrates. The liver fluke is a well-known parasite found in the bile passage of the liver of sheep and cattle, where it inflicts fatal damage. This fluke has a complex life cycle that involves several hosts. The eggs leave the vertebrate host in the feces and hatch in water into free-swimming larvae called miracidial. These larvae swim until they encounter a snail—the intermediate host—into which they burrow. Inside this host the larvae multiply and eventually leave the snail as another larval form called cercaria. Up to 600 cercariae may be produced from a single miracidium. These then enclose themselves within cysts on grass, where they are eaten by the vertebrate host, in which they develop into sexually mature flukes.

Tapeworms

The adults of this parasitic group are found as internal parasites in the digestive tract of vertebrates. Like the flukes, tapeworms have no obvious sense organs, and their complex life cycle usually involves two more hosts. The beef tapeworm, for example, has two hosts—cattle and man.

Tapeworms need no digestive system because they are surrounded by digested food in the gut of their host, and they simply absorb nutrients through their skin. These worms consist of a head, or scolex. Segments, or proglottids, bud off from the neck. The head usually bears hooks and suckers for attachment to the gut lining of the host. The proglottids remain attached in a long chain as they mature, finally breaking off and leaving the host in the feces. The eggs hatch from the proglottids once they have been eaten by the intermediate host.

The candy-striped flatworm is a marine member of the order Polycladida. The order name refers to the many-branched intestinal system with which the organism digests its food. It reaches 2 inches (5 centimeters) in length. Gland cells below the epidermis secrete an adhesive substance that allows the flatworm to stick to a substrate or to prey. The combination of this substance with beating cilia and muscular contractions enables the flatworm to move over any surface.

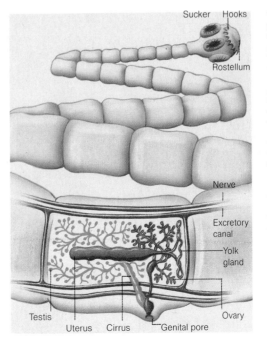

Sucker   Hooks
Rostellum
Nerve
Excretory canal
Yolk gland
Testis   Uterus   Cirrus   Genital pore   Ovary

The reproductive segment (proglottid) of a tapeworm contains male and female reproductive organs. During copulation, the cirrus of the male is inserted into the genital pore of another proglottid, and the fertilized eggs are stored in the uterus. Cross-fertilization usually occurs with a proglottid from another worm in the same host or with another proglottid on the same worm if the worm has twisted around itself. The proglottid breaks off the parent organism, and when the eggs are mature the proglottid wall ruptures, releasing them.

Nematodes and annelids

The nematode parasitizes the intestines of humans, cattle, and pigs, as well as other invertebrates. The adult worms breed in the intestine, and the eggs develop in cysts in the feces. When this excrement is taken in by another host, the eggs hatch in the small intestine and break through the wall. Once in the bloodstream, the young worms are carried to the lungs, make their way to the mouth, and are swallowed. They then burrow into skeletal muscle, remaining there until the flesh containing them is eaten by another host.

In the evolution of metazoans, the stage of development after that of the simple three-layered structure of the flatworms seems to be the appearance of a coelom, or fluid-filled cavity between the body wall and the gut. Another feature that appears at this stage is metamerism, or segmentation, displayed particularly in the annelids. All of these animals share some features but are too different to be placed under one phylum, so each has its own.

Ribbon worms

This group, known as nemerteans, do not have a true coelom. Instead, a solid tissue called a parenchyma fills the cavity between the gut and the body wall. In this respect nemerteans resemble the flatworms. But nemerteans are more highly specialized, with more complex nervous and circulatory systems, and an alimentary canal with a mouth and anus.

Most of the 900 or so species of nemerteans are marine burrowers and feed on invertebrates. They have a remarkable food-catching tube called the proboscis, which is shot out by water pressure through a pore in the head. In many species the proboscis simply coils around the prey but in some it has teeth or a spine that injects poison into its prey.

Roundworms

This group, the nematodes, found in most environments, and millions may occur in only a couple of acres of soil. Scientists have discovered between 10,000 and 20,000 species of nematodes, and some estimate that tens of thousands more remain undiscovered.

Nematodes have cylindrical, tapered bodies that are covered in a thick protein layer, called the cuticle, which is shed periodically as the animal grows. Water flows in and out of the cuticle and body wall continuously.

The intestines of these organisms run from the mouth to the anus and are enclosed by longitudinal muscles along the length of the body. Roundworms move like snakes by contracting these muscles, helped by the elasticity of the cuticle and the pressure of the fluid in the pseudocoel. They have a simple nervous system, as do most aschelminths. The brain is located at the front end of the body and is made up of a ring of ganglia (bunches of nerve tissue), from which nerves run down the length of the body.

Nematodes have separate sexes and give birth to larvae that resemble the adult. Many are parasitic—some only in the larval stage, others only when they are adult, and still others are parasitic throughout their life. Some parasitic nematodes damage crops by sucking the contents from punctured plant cells or by feeding on the tissues inside the plant.

Nematodes also parasitize animals, including humans. The hookworm does great harm by feeding on the blood and cells of the intestinal lining; female hookworms produce many eggs, which leave the host's body in the feces. In unsanitary conditions, the eggs hatch, and the young hookworms enter the human body by boring through the skin of the feet.

Annelids

Annelids, members of the phylum Annelida, are worms whose body is divided into many segments. As such, they are a little more advanced than the unsegmented nemerteans and nemotodes. The phylum Annelida contains about 12,000 species divided into three classes: Polychaeta, Oligochaeta, and Hirudinea. In the polychaete ragworm each segment of the body—apart from the head and the last segment—is identical, and the external and internal organs are repeated in each segment. In the oligochaetes, such as the earthworms, and the Hirundinea, or leeches, the segments are not all identical and some are

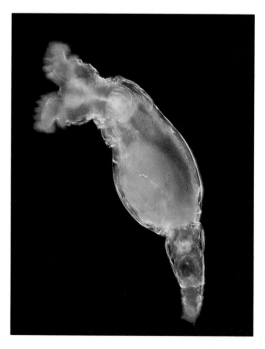

Adult worms · Eggs · Host · Host ingesting feces · Hatching nematode · Small intestine · Skeletal muscle · Host

Rotifers have a crown of cilia around the head region that beat very fast to propel it. In some species this cilia ring looks like a spinning wheel. The species shown here retracts its corona as it creeps along and uses the two spurs at its foot as a means of attachment. The white oval shape in the head region is the mastax, or pharynx, which is used to catch and break up food. The orange area is the stomach. This species reproduces only asexually and has no males.

specialized for particular functions. Annelids have a true coelom.

About 70 per cent of all annelids belong to the class Polychaeta. Almost all polychaete annelids are marine. Carnivorous polychaetes typically have a well-developed head region with eyes, sensory tentacles, and powerful jaws. They move by using the fleshy, leglike extensions of the body, called parapodia, on each segment. Each parapodium has many stout bristles called setae. Some polychaetes, such as the ragworm, are fast-moving predators. Others, such as lobworms, burrow in mud, while still others, such as the fan worms, live in tubes.

Most of the 3,000 or more species of oligochaetes are burrowers and live in freshwater or land habitats. Oligochaetes lack, and they have fewer setae than polychaete. Oligochaetes worms travel by peristaltic contractions, as do burrowing polychaetes, using the bristles as a temporary anchor. The familiar earthworm is an oligochaete.

Polychaetes and oligochaetes reproduce sexually. However, polychaetes occur in separate male and female form, while oligochaetes are hermaphroditic, each worm having both male and female sex organs. The sex organs are located toward the front end of the animal in different segments. During copulation earthworms come together in opposite directions with their undersides pressed together. They are held together by a tube of mucus that is secreted by the glands in the clitellum, an enlarged ringlike segment on the body. The clitellum of each worm attaches itself to the segments containing the spermathecae of the other worm. Sperm is moved to the clitellum by muscular contractions and passes through a groove in the clitellum into the spermathecal opening of the opposite worm. After a few days, the clitellum secretes a hard ring of material, which slips forward and collects eggs and sperm as it moves over the genital openings. When it comes off the worm, the ends seal up. This cocoon can carry 20 eggs, which hatch after about 12 weeks.

Leeches are found in fresh water, the sea, and on land. They share several features with oligochaetes, including hermaphroditism, fertilization in a cocoon secreted by a clitellum, and the lack of parapodia. However, all but one species lacks a coelom. Most leeches are parasitic.

Leeches have flattened, muscular bodies with a sucker at each end, which is used as an anchor while they move by extending and contracting the body. Most leeches feed on the surface of their host. The front sucker is clamped onto the prey's skin, and the teeth in the leech's mouth then make a small cut, which is unnoticed by the host due to an anesthetic that is secreted by the leech. While the leech sucks blood out of the wound, it secretes a chemical called hirudin, which prevents the victim's blood from clotting. Bloodsucking leeches feed infrequently, but when they do feed they can draw out several times their own weight in blood in one meal.

This species of tube-worm inhabits the Mediterranean Sea. It lives in a tube that is made of sand cemented by an organic material that it secretes from special glands. The tube is constructed on the seabed. The movement of small cilia on the tentacles of the worm creates water currents, which carry food toward its mouth.

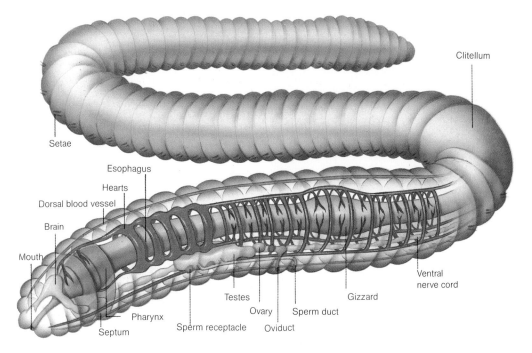

Setae

Esophagus

Hearts

Dorsal blood vessel

Brain

Mouth

Pharynx

Septum

Testes

Sperm receptacle

Ovary

Oviduct

Sperm duct

Gizzard

Ventral nerve cord

Clitellum

Earthworms diffuse gases and water through their cuticle (skin), although oxygen is also distributed by hemoglobin in the blood. These animals have five hearts from which branched vessels carry blood along the length of the body. The nervous system of earthworms is well developed compared with that of the flatworms: clusters of cerebral ganglia form a brain from which two nerve cords run, and although the worm has no eyes, it has light sensitive cells in the epidermis, particularly at its front and back ends.

Echinoderms

The phylum Echinodermata ("spiny-skins") consists of about 6,000 species of exclusively marine organisms, which include such well-known animals as starfish and sea urchins. The description "spiny-skins" is derived from the limy plates that form the endoskeleton of these animals, which in some species is covered with spines. These spines are particularly prominent in sea urchins.

The most obvious feature of these animals, however, is their five-rayed radial symmetry. Zoologists believe that this characteristic, far from primitive, developed at a later stage in echinoderm evolution because fossils of extinct species show that some were bilaterally symmetrical. The fact that echinoderm larvae are bilaterally symmetrical is further evidence of this evolutionary change.

The relationship between the echinoderms and other invertebrate phyla is somewhat obscure. It was initially thought that echinoderms and hemichordates (primitive chordates) were closely related because of the similarity between the larvae of the two groups. It is now believed, however, that the resemblance is due not to genetically similar characteristics but to the adaptation of different organs to perform the same function, a process that is known as convergent evolution.

Sea stars

The 1,600 species of starfish, more properly called sea stars (subclass Asteroidea), are the most familiar echinoderms. They consist of a central disk, from which five arms arise, though some species have more than five arms. These animals move using tube feet called podia, which are found in a groove under each arm. Each podium ends in a suction disk, which allows the sea star to stick to rocky surfaces. When moving over sand, however, the sea star uses its tube feet as stiff legs.

Sea stars also use their tube feet to capture and hold prey, which usually consists of clams and oysters. The tube feet exert enough pressure to open them slightly so that the sea star can push its stomach, which it can turn inside out, into the gap—the stomach can squeeze through a space as small as .004 inch (0.1 millimeters). The stomach then secretes enzymes that slowly digest and absorb the victim. The sea star intestine has five branches, one in each arm. Indigestible fragments are usually ejected via the mouth.

Within each arm there is also a pair of sex organs. Reproduction in most echinoderms is a simple process. The sexes are separate, and sperm and eggs are released into the water, where fertilization occurs. The larvae that hatch are called bipinnariae and metamorphose gradually into the adult form. A few species copulate, and the eggs develop in a brood chamber without going through a larval stage. Sea stars are also able to reproduce by fragmentation—or regeneration, as this process is sometimes called—but this happens only occasionally.

In large numbers, sea stars can have a devastating ecological effect. The population of the crown-of-thorns sea star, for example, has increased in the Pacific Ocean in recent years, and, because it feeds on the live corals, has destroyed whole areas of coral reef.

Brittle stars

The 2,000 species of brittle stars (subclass Ophiuroidea), which occur at all depths of the ocean, are easily distinguished from sea stars by their round central disk and their very long arms. They also differ from sea stars in several other ways. Brittle stars move not by means of tube feet but by muscular movements of the arms. Most brittle stars feed on small living or dead organisms, but do not have an intestine or anus, and indigestible fragments are ejected through the mouth. Sea stars and sea urchins breathe by means of skin gills, which are located all over the body surface, whereas brittle stars use respiratory pouches, called bursae, which occur near the arm bases.

Sea urchins and sand dollars

Sea urchins belong to the class Echinoidea,

The water vascular system of echinoderms is an arrangement of fluid-filled canals. A porous structure called a madreporite links the upper side with a ring canal near the underside. Radial canals run from this water ring along the underside of each arm and give rise to lateral canals, which end in a bulbous ampulla; this itself ends in one or more tube feet (podia). When the ampulla contracts, fluid is pushed into the podium and elongates it. As it touches a surface, the center of the tip of the podium contracts to form a vacuum and sticks to the surface.

Ampulla — Madreporite — Polian vesicle — Ring canal — Tiedemann's body — Radial canal — Lateral canal — Tube feet

The brittle star lives in dense communities on muddy, gravel surfaces. It is a small animal, with a disk only about one-half inch (1.25 centimeters) to just over one inch (2.5 centimeters) across. Brittle stars are the most mobile echinoderms. They move by lifting their disk off the bottom, extending one or two arms forward, and trailing two behind, while the lateral arms push against the bottom with the aid of spines. This species has hooked spines on the underside, which provide additional traction during movement.

which has about 1,000 members. Other members of this class include heart urchins and sand dollars. Most sea urchins have a spherical body, with the skeletal plates fused together to form a hard shell called the test. In most species, the body is further protected by sharp spines, which may be poisonous. Some of the spines are movable, attached to the test in sockets, and are used for walking. The spines of some species of urchin are also used for boring holes in coral or rock, into which the animals wedge themselves to prevent removal. In addition to spines, urchins and sea stars have tiny pincerlike structures on stalks, called pedicellariae, on the skin surface. They use the pedicellariae for defense, for catching small prey, and for cleaning their body surface.

The tube feet of echinoids protrude through holes in the skeletal plates in 10 rows. In heart urchins and sand dollars, the podia are modified for respiration, and they move by means of the spines only. These urchins also burrow in sand. As a result, the heart urchin has reduced spines to assist burrowing, giving it a furry appearance, and the sand dollar's body is covered with tiny movable spines used for crawling and digging.

Sea urchins are omnivorous, often scavenging on organic debris. The mouth is on the underside, and the anus is on the upper side.

Sea cucumbers

As their name suggests, sea cucumbers (class Holothuroidea) are cucumber-shaped echinoderms. They have soft bodies covered with a glandular skin, and their skeletal plates have been reduced to microscopic bony structures.

Sea cucumbers burrow in sand. To breathe, they have tubes known as "respiratory trees," which carry seawater into the body from the anus, so that gaseous exchange can occur internally. They also use their podia and skin surface for breathing.

Sea lilies and feather stars

Sea lilies and feather stars (class Crinoidea) are among the most primitive of echinoderms.

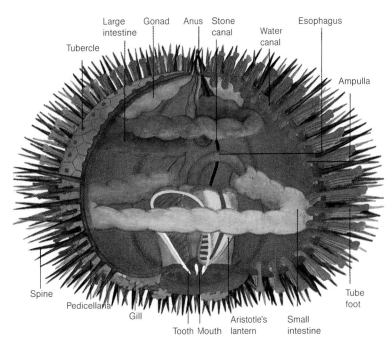

Paleontologists have identified crinoid fossils dating back 600 million years. The 550 species of crinoids basically resemble an inverted sea star.

In the case of the sea lily, this starlike structure is attached to the seabed by a stalk. Most fossil echinoderms are of this form. The sea lily skeleton, which stands on the stalk, is cup-shaped with arms. Tube feet are present in a groove on the upper surface of the arms, but they are used solely for respiration. The arms collect food, usually plankton, and cilia direct it to the mouth on the upper surface. Sea lilies are deepwater creatures.

Feather stars resemble sea lilies, differing, however, in that they are free-living as adults, either creeping along the seabed or swimming.

Sea urchins have a unique mouth consisting of five pyramid-shaped plates that work together by means of muscular action to scrape up food. The mouth is called an Aristotle's lantern because it resembles an old-fashioned lamp. From the mouth, the esophagus ascends and joins the small intestine, which does a complete circuit inside the urchin; but when it joins the large intestine it does another circuit in the opposite direction before it joins the rectum.

The sea cucumber ranges from 4 inches (10 centimeters) to 3 feet (91 centimeters) in length. In the region of the mouth the tube feet have been modified to form tentacles, which are used for food gathering. This species feeds on plankton, which it sifts through its feathery tentacles.

Mollusks

More than 100,000 species of the phylum Mollusca are known to exist, and scientists continue to find many new species every year. In addition, the fossils of about 100,000 extinct species have been found. The three largest molluskan classes are the Gastropoda, which consists of snails, slugs, and limpets; Bivalvia, which includes oysters, mussels, and clams; and Cephalopoda, containing squids, cuttlefish, and octopuses. Mollusk fossils have been found in rocks about 550 million years old and are often well preserved because of their hard shells. But despite their age and the intact fossils, scientists still do not fully understand the origin of mollusks and their relationship to other invertebrates.

Mollusk anatomy

The three main mollusk classes look very different, but all follow the same basic plan. The main bulk of the body contains the internal organs and is called the visceral mass. All mollusks have a skinlike organ called a mantle—a fleshy extension of the body wall—which hangs down on each side of the visceral mass. The space between the mantle and the visceral mass is called the mantle cavity. The mantle secretes the shell, which is present in most mollusks. In mollusks with no outside shell, the mantle forms a tough cover around the body organs.

Almost all aquatic mollusks breathe with the aid of comblike gills in the mantle cavity called ctenidia, which are covered with many small, hairlike structures called cilia, whose rhythmic movement draws water over the gill surface. Blood that flows through the gill filaments takes up oxygen from the water current, and carbon dioxide from the blood diffuses out of the gill filaments.

The kidneys and anus open into the mantle cavity, and waste is also carried away by the water the mollusk exhales. In some species, the edges of the mantle may be joined to form tubes called siphons, which create a one-way flow of water through the mantle cavity. Many mollusks also have sensory cells on the edge of their gill membrane. These organs, called osphradia, are thought to be sensitive to chemical signals and able to detect the juices of prey or other food sources. They also determine the level of sediment in the incoming water, too much of which would block the delicate gill filaments.

Gastropods

Ancient gastropods had a mouth and anus at opposite ends of the body, as do the larvae of present-day gastropods. But early in the development of larval gastropods a remarkable change takes place—the visceral mass twists through 180°. This process, known as torsion, is important because it brings the mantle cavity from a lower backside position to an upper frontal position. The shell then only needs one opening, and the mantle cavity provides a space into which the young gastropod can withdraw for protection.

The problem of fouling by waste released over the head is avoided by having holes in the shell that direct excrement away from the head (as is found in keyhole limpets and the abalone), or by shifting the anus to the side. A one-way water flow through the mantle cavity also helps avoid this problem.

The left and right sides of the gastropod body grow at different rates because of the spiral shape of the shell. The organs on one side do not develop, and the visceral mass, mantle, and shell become spirally coiled. This development also makes the long digestive system of the gastropod more compact.

Most aquatic gastropods breathe by means of gills, but land-dwelling snails and slugs (subclass Pulmonata) lack gills and have modified the mantle cavity into a lung. Air is moved in and out of the mantle cavity by raising and lowering the mantle, which is moist and has a rich blood supply—essential features for gaseous exchange to take place.

Pulmonates show another adaptation for a land-dwelling existence in that nitrogen-containing waste is excreted as uric acid crystals. Aquatic gastropods, in contrast, excrete nitrogen-containing waste as ammonia, which is toxic in concentrated form and therefore re-

The common garden snail is typical of terrestrial pulmonates and shows clearly how the gills that lie in the mantle cavity of aquatic gastropods have evolved into a lung in land-dwelling gastropods. The visceral mass becomes twisted during torsion, but in addition, the anal and pulmonary openings come to lie next to each other at a single opening in the shell, called the pneumostome. The position of the pneumostome at the side of and behind the head helps to prevent fouling.

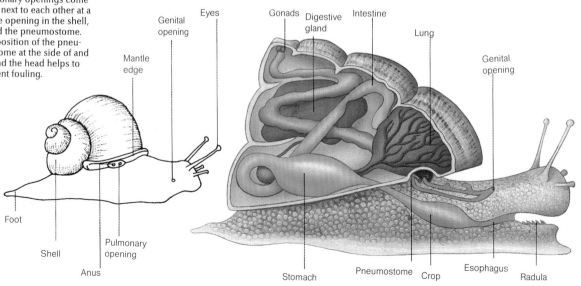

Foot

Shell

Anus

Pulmonary opening

Mantle edge

Genital opening

Eyes

Gonads

Digestive gland

Intestine

Lung

Genital opening

Stomach

Pneumostome

Crop

Esophagus

Radula

quires a large amount of water to dilute it.

Gastropods are generally considered to be inhabitants of damp places, but the terrestrial adaptations and, more importantly, the possession of a shell, into which the animal can withdraw and seal itself off from the environment, have enabled them to invade such unlikely regions as deserts. Some pulmonates have secondarily returned to an aquatic existence, but they still breathe by means of a lung.

Most gastropods move on a flat sole, which in many species has a large pedal gland, which secretes mucus onto the surface over which the sole moves. Locomotion is achieved by waves of muscular contractions that pass down the foot. In the sea hare, the foot is formed into folds, or parapodia, which may be used for swimming. In the sea butterflies, or Pteropoda, the parapodia are drawn out into winglike structures.

Most gastropods feed using a radula—a serrated band of teeth—to rasp off fine particles from their food. Many species, such as land snails and limpets, feed on vegetation. Others, such as whelks, are carnivorous. They detect their prey by means of a structure that is sensitive to both chemical signals and touch, called an osphradium. The mouth of carnivorous gastropods often lies on an extension of the head, called the proboscis. The radula of these species has fewer, larger teeth than the plant-feeding gastropods. In the cone shells it has become a stalklike stinger with which they stab and inject a poison into their prey. Some carnivorous gastropods eat oysters. They do so by secreting a chemical that softens the oyster shell, then using the radula to wear it away. By alternately softening and rasping the shell, the carnivore makes a hole in the shell, through which it can insert the proboscis to eat the oyster.

Aquatic gastropods have separate sexes, but land snails and slugs are hermaphroditic. Male gastropods either shed sperm into the water, where they fertilize eggs from the females, or deposit them into the female with a penis. The female gastropod lays eggs singly, in strings, or in a thin shell. The females of some species carry their eggs in their bodies until they hatch.

In some gastropods, the young pass through two stages: the first is the "trochophore" stage, in which the larva is roughly spherical with a band of cilia for movement. It later develops a shell to become the veliger larva. In other gastropods, the trochophore stage remains in the egg, which hatches to produce the veliger, and in some species, both the trochophore and veliger stages remain in the egg.

Some aquatic snails are hosts to human platyhelminth parasites, such as the liver fluke and the blood fluke, the cause of the disease bilharzia.

Sea slugs are gastropods which, in secondary adaptation, have become untwisted so that the mouth and anus are at opposite ends of the body. The cluster of appendages on this species are secondary gills arranged around the anus. Some sea slugs are protected by the stinging cells of their jellyfish prey, which settle in the extensions of the sea slug's body.

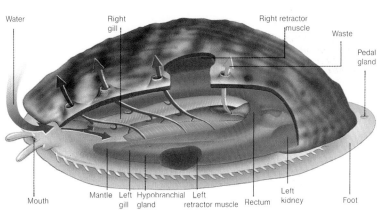

Water   Right gill   Right retractor muscle   Waste   Pedal gland

Mouth   Mantle   Left gill   Hypobranchial gland   Left retractor muscle   Rectum   Left kidney   Foot

The abalone, or ormer, is a primitive gastropod that lives in shallow water. Its broad shell is asymmetrical with a single spiral and several holes. Incoming water bathes the gills. Blood flowing through the gills absorbs oxygen from the water and releases carbon dioxide. The water that the abalone exhales through the shell holes removes carbon dioxide and waste.

The queen scallop is a bivalve mollusk. It swims by jet propulsion, clapping its valves together, but usually only when disturbed. During the evolution of mollusks, these animals lost their head, with the result that all their sense organs are found on the edge of the mantle. These organs consist of eyes and tentacles that carry cells that are sensitive to touch and chemicals in the water. Scallops filter-feed, and breathe through W-shaped ciliated gills, which direct a complex pattern of water currents through the body.

Bivalves

The class Bivalvia includes clams, oysters, and mussels, and is characterized by the possession of a shell that is divided into two halves joined by an elastic ligament.

Bivalves are mainly marine, although some live in fresh water. The foot is adapted for burrowing mainly in mud and sand, but some species can bore into hard material. The shipworm, for example, drills into wood using the roughened edges of its shell, and other species even bore into rocks.

The body of a bivalve is long and laterally flattened, surrounded on each side by two lobes of the mantle. Each lobe secretes a shell called a valve. The gills hang from the roof of the mantle cavity and lie on each side of the body. They are used for collecting food as well as for breathing. Bivalves have no head—it would be impossible to perceive stimuli while buried in the sand—so the sense organs occur on the edges of the mantle, and these detect light and test water currents.

Most bivalves are filter-feeders, straining very small food particles from the water. The gills are highly modified for this purpose and secrete mucus in which food particles are trapped. A complex arrangement of cilia draws water into the mantle cavity, sorts the particles, and carries food to the mouth. Flesh folds near the mouth (palps) sort the trapped particles, and nonfood material is passed out with water while food particles pass into the stomach. There, a gelatinous rod called the crystalline style is rotated by cilia and rubs against the stomach wall and style sac, releasing enzymes that partly digest the food particles. Further digestion occurs within the cells of the digestive glands that surround the stomach.

Most bivalves use their foot as a burrowing organ. The two valves close and water is pushed out of the mantle cavity, which loosens the mud and makes the movement of the foot easier. Some species have serrated shell edges that also break up the mud and facilitate burrowing. Blood is then pumped into irregular channels, called blood sinuses, in the tissue of the foot. This action makes the top of the foot swell and anchors the animal. Contractions of the foot muscles then pull the animal down.

Most bivalves are sluggish, but some, such as the razor clam, have a thin, streamlined shell and large foot and can burrow very rapidly. Scallops use jet propulsion to swim by clapping the two valves of the shell together. Other species, such as oysters, which cement themselves to rocks, and mussels, which attach themselves to surfaces by strong threads of organic material, are unable to move.

In most bivalves, males and females shed sperm and eggs into the water, where fertilization produces a trochophore larva, which develops two valves to form the veliger larva. The veliger swims around in the open sea for approximately two weeks before settling on the bottom to become an adult. In the freshwater mussel, however, the female carries

The squid is carnivorous, feeding mainly on fish and crustaceans. It locates its prey with its highly evolved eyes, darts toward it by rapidly ejecting water from the mantle cavity, and seizes the prey with its tentacles. The squid pulls prey to the mouth and holds it with its arms while it uses its powerful horny jaws to bite off the prey's head. The squid tears large chunks off the prey, pulls them back with the radula, and swallows them, rejecting only the gut and tail.

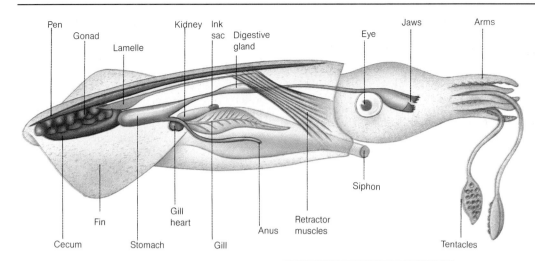

Pen · Gonad · Lamelle · Kidney · Ink sac · Digestive gland · Eye · Jaws · Arms · Siphon · Cecum · Fin · Gill heart · Stomach · Gill · Anus · Retractor muscles · Tentacles

Squids, like all cephalopods, are highly adapted to a deep-sea, carnivorous existence. They use the water that flows through them for locomotion and as a source of oxygen. Squids can cruise or hover as well as dart, using their fins as stabilizers. They have 10 arms and 2 tentacles with which to catch prey. Food is absorbed not in the digestive gland, as in other mollusks, but in the cecum.

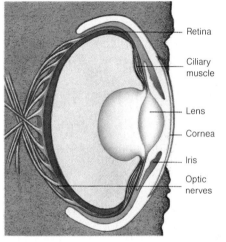

Retina · Ciliary muscle · Lens · Cornea · Iris · Optic nerves

The cephalopod eye is highly developed and resembles the vertebrate eye in structure. The retina contains light-sensitive bodies, which are pointed toward the light source. An image is formed, but how clear it is is not known. Cephalopods also have color vision.

eggs on her gills. Sperm enter the mantle cavity with the water currents and fertilize the eggs there. The larvae then leave the mantle cavity and live as parasites on the gills of fish until they develop into adults.

Cephalopods

The cephalopods represent the highest stage of molluskan evolution. The three main groups belong to the subclass Coleoidea and are the squids (order Teuthoidea), the cuttlefish (order Sepioidea), and the octopus (order Octopoda).

All cephalopods are adapted to a free-swimming life in the sea, but only one living species—the nautilus—has an outer shell. Most other cephalopods have an inner chambered shell, which varies from species to species. In the cuttlefish, it is called a cuttlebone, and in the squid, a pen. The shell lies along the dorsal parts of the animal's body and is mostly filled with gas, which provides lift in water. Toward the tail end, however, the shell is filled with liquid, which adds weight to that part.

In the cuttlefish and the squid, the shell has been reduced to a small plate, and their bodies are elongated and streamlined. The octopus, however, is less streamlined and lacks a shell. The edges of its mantle are fused together, forming a bag around the body, from which the head protrudes, surrounded by tentacles. The octopus usually crawls around the seabed using the suckers on its tentacles, but it is capable of jet propulsion when fast movement is necessary. The octopus propels itself by pumping water into the mantle cavity and forcing it out through a funnel below the head.

All cephalopods are carnivorous. Cuttlefish and squid catch their prey with two long tentacles with suckers at the tips, and with arms—8 in the cuttlefish and 10 in the squid—that are equipped with suckers all along their length. The octopus has 8 tentacles of equal length, any one of which can be used to grab prey. The octopus bites its prey with its two horny jaws, and its poisonous saliva then enters the wound and paralyzes the prey. Large pieces of flesh are torn off, pulled into the mouth by the radula, and swallowed.

Cephalopods have the most highly developed brains of all invertebrates. Squid and octopus can learn and remember information. Most cephalopods rely on their good eyesight to detect food and enemies. In addition, sen-

sory cells on the octopus's tentacles detect minute concentrations of chemicals, which also alert it to the presence of prey or to possible danger.

Many cephalopods can camouflage themselves in seconds, changing color by the expansion or contraction of small bags of pigment in the skin. Some species also have an ink sac—a pouch off the intestine—which releases a cloud of black liquid through the anus when the animal is alarmed, behind which it can make its escape.

Male cephalopods produce sperm that are enclosed in a case called the spermatophore. One of the male's arms is modified to transfer the spermatophore into the mantle cavity of the female, where the eggs are fertilized. The female may stick them onto rocks, where they develop into adults without passing through a trochophore or veliger stage.

Introduction to arthropods

The phylum Arthropoda contains more than three-fourths of all the different species of animals. It includes crustaceans, spiders, and insects, as well as many smaller groups. Among the most important groups of arthropods are insects, crustaceans, arachnids, centipedes, horseshoe crabs, and millipedes. Insects make up the largest class of arthropods in terms of the number of species.

An arthropod's exoskeleton, which is jointed to allow movement, is hardened and does not grow with the animal. As a result, it must be periodically shed—a process known as molting—in order to allow growth. The arthropod body is usually divided into head, thorax, and abdomen, although in some groups there may be no clear distinction between the three regions. The head and thorax may be fused, forming a cephalothorax, or the abdomen may be reduced in size. The head typically carries feeding and sensory structures. Each segment of the body usually has a pair of jointed appendages, which are modified for specific functions in different species.

Arthropods have a body cavity (coelom), but it is small and contains only the gonads and excretory organs. The other internal cavities form a hemocoele (blood cavity). The circulatory system is open, with the heart in the upper part of the body. The digestive system consists of a tubular gut that runs from the mouth to the anus. The foregut and hindgut are lined with chitin and are shed during molting. Excretion is through specialized tubes and the anus is at the back end of the body.

Arthropods have a nervous system with a brain and a nerve cord, which has branches called ganglia in each segment. Bristles that are sensitive to touch are also a common feature. Eyes, which may be simple or compound, are usually present. Compound eyes are unique to arthropods and are ideal for detecting motion.

Trilobites

The trilobites are a fossil group of marine arthropods of the class Trilobita that were once extremely numerous but became extinct toward the end of the Paleozoic era, about 230 million years ago.

Most trilobites were less than 4 inches (10 centimeters) long, and the body was divided into a head, or cephalon; a thorax; and a tail, or pygidium. The cephalon was composed of four or five fused segments and carried a pair of antennae, a mouth, a pair of eyes, and four pairs of forked appendages, which functioned as both gills and legs. The segments of the thorax were movable, whereas those of the pygidium were fused to form a solid shield. Each segment of the thorax possessed a pair of walking legs.

Horseshoe crabs

Despite their name, horseshoe crabs (order Xiphosura) are not crabs at all, but primitive marine arthropods. They are among the only surviving members of a large group of ancient animals of the class Merostomata.

The horseshoe crab's body is divided into two parts: the prosoma, which is covered by the shell and includes the head, and the abdomen. On the underside of the prosoma is a mouth, with a pair of pincerlike feeding appendages called chelicerae, and five pairs of walking legs. There are five pairs of abdominal appendages which are modified as leaflike book gills. The book gills keep a constant current of water circulating over the gills and also act as paddles, which move the animal. The horseshoe crab's nervous system is well developed, but the eyes are not. They can detect movement but cannot form an image.

The upper exoskeleton of trilobites was much thicker than the one on the underside, which is why most fossils display the upper view. The name Trilobita derives from the triple-segmented transverse divisions that ran down the animals' length. Trilobites were bottom dwellers and scavenged for food.

This species of horseshoe crab is found in the shallow coastal waters of Asia, the Gulf of Mexico, the west coast of North America, and the northern Atlantic. Its telson, or tail, which it uses to push and dig with, is also used to make threatening gestures, as in this position.

During mating, the female horseshoe crab carries the male on her back to shore, where she digs several holes and lays from 200 to 1,000 eggs in each. The male then fertilizes the eggs. Newly hatched horseshoe crabs are called trilobite larvae because of their similarity to the larvae of that class.

Sea spiders

The class Pycnogonida contains about 600 species of carnivorous marine animals known as sea spiders. They feed on corals and sponges and are found in all marine waters. The sea spider's head bears clawlike structures called chelicerae, which are used for seizing and tearing apart food; a sucking mouth at the end of a tube; and four eyes. Despite their common name pycnogonids bear no real similarity to true spiders. Pycnogonids have a segmented abdomen and legs that end in claws, whereas spiders do not have these characteristics.

Centipedes and millipedes

The animals in the class Chilopoda are commonly known as centipedes, and those in class Diplopoda as millipedes. Millipedes and centipedes are found mainly in damp conditions, such as rotting logs or in leaf litter, because they do not possess a waxy cuticle with which to reduce water loss.

Both groups have separate sexes, and the female lays eggs that are fertilized by the male. Some species lay the eggs in a "nest," where they are guarded by the female, but others, such as the centipede *Lithobius,* lay one egg at a time and then leave it. The young resemble the adult and, in some species of centipedes, have the same number of segments. But other young centipedes and all millipedes have fewer body segments than do the adults.

Centipedes are fast-moving carnivores. They have one pair of antennae on the head and two pairs of jaws. The eyes of centipedes may be simple ocelli, or clusters of light-sensitive cells, or modified compound eyes. Centipedes are flattened and divided into a large number of segments. All of these, except for the first, carry a pair of long, slender, walking legs.

Millipedes have a lower "lip" called a gnathochilarium. Many millipedes lack eyes completely, and those that have eyes have only ocelli. Most millipedes are vegetarian scavengers. Their body is also segmented, but is usually cylindrical. The first four trunk segments differ from the others in that the first is legless, and the following three bear only one pair of walking legs. These four segments together are sometimes called the thorax. The abdomen has many segments, which are fused together in pairs called diplosegments. Each diplosegment bears two pairs of short legs, and as a result, a millipede may have as many as 115 pairs of legs. Nevertheless, millipedes are not good runners and, when attacked, they protect themselves by coiling up. Many species also emit a foul-smelling secretion from special stink glands on the sides of the body.

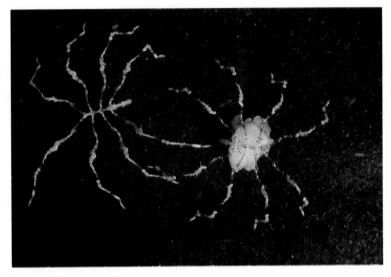

Pycnogonids, or sea spiders *(above),* are notable for their pair of special legs, called ovigers, at the rear of the head, which they use to groom themselves. Males use their well-developed ovigers to carry around masses of fertilized eggs until they hatch.

Centipedes are carnivorous and kill their prey with a large pair of poison claws called maxillipeds, found in the first body segment. Each claw has a pointed fang that is fed by a poison gland. This cave centipede is eating a cricket.

Crustaceans

The crustaceans became the major group of marine arthropods when the trilobites and giant sea scorpions (Eurypterida) became extinct. The group, called Crustacea, contains about 30,000 species and includes such animals as water fleas, barnacles, crabs, lobsters, shrimps, and wood lice (pill bugs). It also includes tiny zooplankton, which live near the surface of the sea and occupy an important position in aquatic food chains. Although Crustacea has traditionally been considered a class within the phylum Arthropoda, some zoologists believe the size of the group means it should be considered a subphylum.

Characteristic features

The exoskeleton of crustaceans, which is composed of chitin and hardened with calcium salts in such species as crabs, is molted periodically to enable the animal to grow. This is an important process in the life of crustaceans and other arthropods, and occurs in a strict sequence of events. First, useful materials contained in the shell, such as calcium salts, are reabsorbed into the body; the new cuticle then begins to form under the old one. The animal swells up, causing the old cuticle to split, and the new cuticle is hardened. Before the new cuticle hardens, the animal is easy prey for predators and usually seeks shelter during this period.

In many crustaceans an outer shell, or carapace, covers the exoskeleton of the thorax or anterior trunk segments. Typically, the body consists of a head, thorax, and an abdomen, but often the head and a number of thoracic segments are fused to form a cephalothorax.

The head region has two pairs of antennae, which is characteristic of crustaceans. It also has one pair of mandibles, which in most species are heavy and have grinding and biting surfaces, and two pairs of maxillae.

The trunk region of the cephalothorax is segmented, the number of segments varying according to the species, and each segment bears a pair of appendages. These appendages are different in each species, being modified to suit particular functions. Each abdominal segment usually also has a pair of appendages structured for various functions, such as swimming, crawling, breathing, capturing food, or reproducing.

The appendages are biramous—that is, each one ends in two jointed branches; they are tubular and jointed and contain muscles that contract to bend the limbs. Crustaceans swim by beating these appendages—some of which have a fringe of bristly setae, to push against the water—and most of the animals also crawl.

Many crustaceans breathe with gills, although some land-dwelling species, such as the robber crab, have modified them to become air chambers lined with blood vessels, which absorb oxygen. The gills also help crustaceans eliminate ammonia waste. The primary organs of excretion are special glands in the head region called the antennal glands and maxillary glands. The crustacean blood system is open—the blood is pumped by the heart into a hemocoele, where it simply bathes the tissues.

The nervous system of these animals is well developed, and the sensory organs include eyes and various receptors sensitive to touch. Most crustaceans have compound eyes in the adult stage. The young, or nauplius larvae, have a median eye composed of three or four clusters of photoreceptors (ocelli), which in some species persist into the adult stage. Other sensory receptors include special touch-sensitive hairs called setae. Setae are scattered over the body surface, but are most concentrated on the appendages and statocysts, or balancing organs.

Crustaceans are usually dioecious—that is, they have separate sexes—but some are hermaphroditic and have both male and female sexual organs. The young hatch from eggs that are usually protected by one of the parents and take the form of free-swimming planktonic larvae that have fewer appendages than the adult. However, through successive molts, trunk segments and extra appendages develop.

Branchiopods

This subclass of mostly freshwater animals have earned their name (meaning "gill feet") from their thoracic appendages, which are modified for respiration as well as filter-feeding and locomotion. The group consists of four orders: Anacostraca, or fairy shrimps, which have no carapace; Notostraca, or tadpole shrimps, whose head and front thoracic region is covered by a carapace; Conchostraca, or clamp shrimps; and Cladocera, or water fleas. The animals in this last group, which include the common genus *Daphnia,* are covered by a carapace that encloses the trunk but not the head. The carapace of these animals is usually transparent, but can appear red or pink, depending on the level of oxygen in the water. *Daphnia,* for

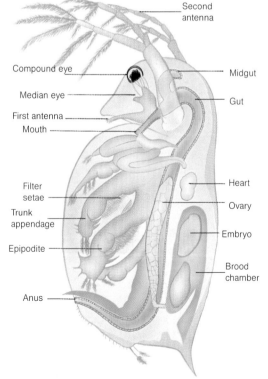

The microscopic *Daphnia,* a genus of water flea, is a typical branchiopod. It has five or six pairs of trunk appendages, which have epipodites—extensions that serve as gills. The appendages also bear setae, which collect food particles. The food is then moved into a food groove and spurts of water push it toward the mouth. Unlike other branchiopods, which use their trunk appendages for locomotion, *Daphnia* uses its large second antennae as paddles to move it in a jerky up-and-down motion. Also, water fleas have a pair of median compound eyes that direct them when swimming.

Second antenna

Compound eye

Median eye

First antenna

Mouth

Filter setae

Trunk appendage

Epipodite

Anus

Midgut

Gut

Heart

Ovary

Embryo

Brood chamber

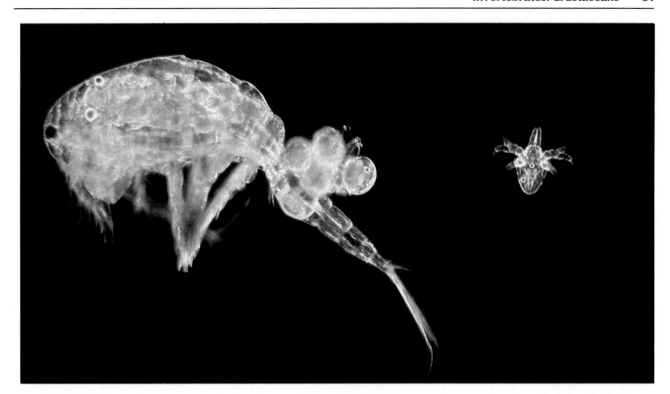

example, is transparent in oxygenated water, but turns pink in stagnant water, when it produces hemoglobin to extract more oxygen.

Branchiopods are filter-feeders that collect food particles on the bristles of the trunk appendages. The food is carried to the mouth by streams of water that pass between the trunk appendages as they move back and forth. The first maxillae finally push the food into the mouth.

The sexes are generally separate, and copulation is the means of fertilization, but under favorable conditions the females reproduce parthenogenetically—that is, the eggs develop without fertilization. The eggs of water fleas hatch into females for several generations until adverse conditions, such as low water temperature or short food supply, induce the production of males, after which eggs are again fertilized by copulation.

Ostracods and copepods

These two subclasses of tiny crustaceans contain freshwater and marine species, and plank-

tonic, or shallow-water, as well as benthic, or deepwater, types. The ostracods are similar to the conchostracan branchiopods in that they have a bivalve hinged carapace. The head is more developed than the remainder of the body, and its appendages (especially the antennules and antennae) are modified for crawling, swimming, and feeding. Most ostracods are filter-feeders, but some species are scavengers, predators, or parasites.

Most copepods live in salt water and occur in huge numbers, making them of great economic and ecological importance because they form a large part of the diet of many fish. The body is short and tubular, with a trunk that is usually composed of six segments, an abdomen, and a reduced head region. The first antennae are long, whereas the second pair is short and often branched. They can move rapidly by beating their thoracic appendages, which causes jerky movements. They are also able to glide slowly by moving the second antennae.

Planktonic copepods usually live in the upper 650 to 975 feet (200 to 300 meters) of the

Cyclops are freshwater copepods. These tiny animals can stagger the development of eggs that are produced from a single mating. After fertilization, one or two egg-containing sacs form on the genital segment of the female—each sac holds up to 50 eggs. A number of eggs are hatched from half a day to five days after fertilization, and a new batch is then brooded. The eggs hatch as nauplius larvae, just as the one shown above, swimming away from the parent.

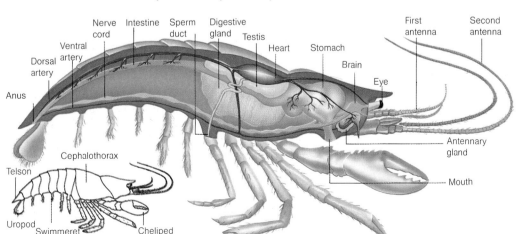

The anatomy of crayfish is similar to that of most decapods, but the male *(left)* and female differ in some respects. For instance, the female has large swimmerets, on which she carries newly hatched larvae, whereas those of the male are small. The male appendages are modified for the transmission of sperm.

Barnacles can be stalked; that is, cemented to a surface by a peduncle, or non-stalked and attached directly to a surface. The peduncle of the stalked goose barnacle carries the capitulum, or body, which is surrounded by hard plates. The upper plate is called the carina.

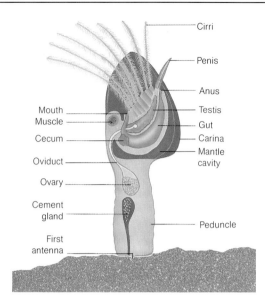

Cirri
Penis
Anus
Mouth
Testis
Muscle
Gut
Cecum
Carina
Mantle
cavity
Oviduct
Ovary
Cement
gland
Peduncle
First
antenna

Cirripedia. The subclass derives its name from the six pairs of featherlike, two-branched legs (cirri) that protrude from the shell and that are used for filter-feeding.

All barnacles are marine, and most are found attached to rocks, shells, or driftwood. Some species, however, live on the bodies of other crustaceans. Barnacles are surrounded by a carapace of calcareous plates, which develops from a clamlike carapace in the young, or cypris, larva.

Most barnacle species are hermaphrodites, but cross-fertilization is common. The young hatch as nauplius larvae, as in other crustaceans, but then develop into cypris larvae that settle on a suitable surface. Finally, metamorphosis takes place, and through a series of molts the animal rapidly adds the carapace and reaches adulthood.

Decapods

The order Decapoda contains most of the larger crustaceans in the subclass Malacostraca, including lobsters, crabs, crayfish, and shrimps. About 10,000 species have been described, and most are marine, although some species live in fresh water. All decapods have eight pairs of thoracic appendages; the rear five are modified for walking and are the origin of the name decapod (five feet). The front three pairs serve as feeding appendages (maxillipeds). In some decapods, the first pair of walking legs is heavier and stronger than the others and has pincers called chelae.

The abdomen is composed of six segments and a tail, or telson. The abdominal appendages are called swimmerets, and in some species they are very small. The sixth abdominal segment usually has a pair of appendages called uropods, which together with the telson form a tail fan.

Decapods have compound eyes on jointed movable stalks, and the central nervous system is well developed. Their wide variation in color usually depends on the habitat, and is caused by the pigment-producing cells called chromatophores in the exoskeleton. Aquatic decapods breathe with gills, usually five pairs,

ocean and commonly migrate to and from the surface daily. Light appears to be the trigger for this behavior, but its value to the animal is far from certain.

Many copepods are filter-feeders, some are predatory, others are scavengers, and still others are parasitic. The parasitic species feed on freshwater and marine fish, some living on the gill filaments, and others in the intestines.

Neither ostracods nor copepods have gills, and gaseous exchange occurs in the valves (in ostracods), or across the body surface (in copepods). Both groups have an open circulatory system. Median eyes are found in both, and only one group of ostracods has compound eyes. One outstanding feature of these animals is their luminescence. In both groups, fertilization is generally by copulation, although parthenogenesis is known in some ostracod species.

Barnacles

Barnacles belong to the only group of crustaceans that are sessile (fixed to one spot): the

The shore crab is one species in a group that includes scavengers and predators of other large invertebrates, such as the lugworm. Shore crabs attack their prey with their pincers and catch them using feeding appendages called maxillipeds, which pass the prey to the other mouthparts. A portion is bitten off, the rest is torn apart, and the morsels are put into the mouth. The heavy claw on the right has blunt serrations on the pincers, which are used for crushing. The lighter claw on the left is developed for cutting.

which run vertically in the cephalothorax between the endoskeleton and the other organs. Water flows in at the base of the front five appendages, bathes the gills, and flows out through vents under the second antennae.

An elaborate courtship ritual is typical of many decapods prior to copulation, and scented chemicals called pheromones play an important role in their sexual behavior. In many species, the eggs are laid soon after copulation and are cemented onto the swimmerets of the female.

Lobsters, like some crabs, have a single huge claw that is used for crushing; the other claw of the pair is much smaller but has sharp edges and is used for seizing and tearing prey. They are scavengers, but also catch fish and break open shelled animals. Crayfish are similar in appearance to lobsters, but most live in fresh water. Also, like lobsters, crayfish are nocturnal and feed on almost any organic matter, living or dead.

Crabs are probably the most successful decapods, in that they can live on land as well as in water. They are found at all depths of water, in all parts of the world. They have a wide carapace and, unlike lobsters, a small abdomen, which is tucked tightly under the cephalothorax. The female uses its swimmerets only for brooding eggs. Crabs can walk forward, but more usually move sideways. The crab's large front claws, called chelipeds, are not used for walking.

Crabs range in size from the pea crabs (Pinnotheridae)—which live in the tubes constructed by marine annelids, on sea urchins, or in the mantle cavity of gastropods—to the Japanese spider crab, whose body measures about one inch (2.5 centimeters) across, and whose leg span is more than 3 feet (1 meter). They exist in a wide diversity of forms, including the hermit crab, which has no shell of its own and takes over empty gastropod shells for protection.

Crabs are filter-feeders, predators, and scavengers, and their method of obtaining food is usually reflected by the shape of their chelipeds.

With their narrow bodies and well-developed abdomens, shrimps and prawns are much better designed for swimming than the lobster, but even so, most of them are bottom-dwellers. Their thoracic legs are usually long and slender, and the first three pairs may have claws. The abdomen has long, fringed swimmerets used for swimming. In females, eggs are attached to them.

Isopoda

This order contains the only group of truly terrestrial crustaceans—the pill bugs—although most species are marine. The shield-shaped body is flat, has no carapace, and has seven pairs of legs. Isopods have a pair of compound eyes and two pairs of antennae.

Water loss can be a problem to wood lice because they do not have a waxy cuticle like some other land-dwelling arthropods, such as insects. They survive by living in fairly damp habitats and by leading a nocturnal existence. Other behavioral adaptations, such as rolling up into a ball, help to reduce water loss.

Arachnids

The class Arachnida is a group of mostly terrestrial arthropods that includes spiders, scorpions, mites, ticks, harvestmen, pseudoscorpions, and sun spiders. Many arachnids have highly developed chelicerae (also found in pycnogonids and horseshoe crabs) in the head region, which usually take the form of a pair of pincerlike organs. It is possible that arachnids evolved from related pincer-bearing arthropods that migrated from the sea to the land. In doing so, arachnids have developed a cuticle, which reduces water loss, and their book gills have become book lungs.

Arachnid anatomy

Like all arthropods, arachnids have a hard exoskeleton made of chitin and jointed appendages, but unlike most of their relatives, true arachnids have no antennae. Their bodies are divided into a cephalothorax and an abdomen. The cephalothorax, called a prosoma, is usually unsegmented, and its upper surface is covered with a carapace. The lower region is usually protected by plates. The abdominal segments of most arachnids, apart from scorpions, are fused. In ticks and mites, both the cephalothorax and abdomen are fused to form a single body.

The first pair of appendages, called chelicerae, are used to grasp prey and to feed. The second pair, called pedipalpi, perform various functions, including grasping, killing, or mating, depending on the species. The remaining four sets of appendages are legs. Although most arachnids are carnivorous, they have no jaws and cannot chew. They feed by secreting or injecting digestive enzymes into the prey, which they then suck in.

The respiratory system of arachnids is made up of specialized breathing organs, called book lungs, and a network of tracheae, which are tubes that carry air from the exterior to the internal organs. Some arachnids have book lungs only, others have tracheae only, and still others have both. The circulatory system is usually open—that is, arteries carry blood from the heart into a series of blood spaces called the hemocoele. Blood is oxygenated as it flows past the book lungs on its way back to the heart. Some arachnids have a copper-based, oxygen-carrying compound called hemocyanin as the major respiratory pigment of the blood. The excretory system includes a pair of Malpighian tubules, which carry wastes from the blood to the gut, and coxal glands, which open to the exterior in the cephalothorax.

The brain consists of two bundles of nerve

The orange garden spider is a typical web-spinner. Like all spiders, it has eight legs joined to the cephalothorax, which also bears a pair of pedipalpi and fanglike chelicerae. Most spiders have eight eyes. Apart from the brain and sucking stomach (which are in the cephalothorax), most major organs are in the abdomen, including the silk glands and their spinnerets.

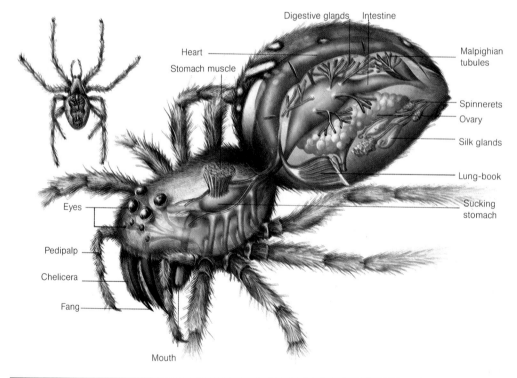

Digestive glands Intestine

Heart

Stomach muscle

Malpighian tubules

Spinnerets

Ovary

Silk glands

Lung-book

Sucking stomach

Eyes

Pedipalp

Chelicera

Fang

Mouth

An orb web is spun in stages and takes a spider about an hour to make. Many species spin a new web every day, usually at night or just before dawn. This may be necessary because the web has been damaged, for instance by the struggles of trapped insects. Rain can damage the webs of some species, although others, particularly in the tropics, build webs strong enough to withstand a storm.

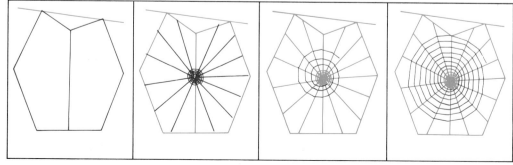

cells, or ganglia, above the esophagus, joined to more ganglia below it. Nerves from the ganglia run to various sensory organs, and there are sensory hairs over the body surface. Most arachnids have simple eyes—spiders usually have eight, scorpions have six to twelve. Even so, their eyesight is, in most cases, very poor.

Reproductive systems vary in arachnids, but the sexes are generally separate. Fertilization is by copulation, usually preceded by an elaborate courtship ritual. Sperm is often transmitted through a packet of sperm called a spermatophore, which is picked up by the female. When hatched, most arachnids are small versions of the adults; they do not metamorphose as do many insects. The young molt several times while growing.

Ticks, their bodies swollen with blood sucked from their host—a European hedgehog—remain attached to the skin of their host and feed off it continuously for several days.

Spiders

There are more than 30,000 known species of spiders. The largest spiders are the tarantulas of South America (suborder Orthognatha), some species of which have a body 3 inches (8 centimeters) across with a leg spread of 7 inches (18 centimeters).

Spiders are all predators and feed mainly on insects. Larger spiders, however, sometimes feed on vertebrates such as fish. Most spiders have poison glands in the cephalothorax, and inject the poison by means of the chelicerae. The poison is used to paralyze or kill prey, or in self-defense. Only a few spiders have venom strong enough to endanger humans.

Many spiders produce silk threads, which they use to spin webs, catch prey, and make cocoons. Abdominal glands secrete the silk through organs called spinnerets. Not all spiders spin webs, however. Trap-door spiders (family Ctenizidae) dig a tunnel, which they line with silk. The tunnel is closed at ground level by a hinged door. When a small animal passes by, the spider jumps out and grabs it. Many wolf spiders (family Lycosidae) and jumping spiders (family Salticidae) do not trap their food but stalk their prey and then leap on it.

Spiders perform a courtship ritual before mating. Male wolf spiders, for example, wave their pedipalpi to attract a female. In many web-spinning spiders, the courting male plucks the threads of the web in a special way so that the female will not mistake him for prey. Before courtship begins, the male spins

a "sperm web," onto which he drops semen, and fills a reservoir at the tip of his pedipalp with sperm. During mating, he inserts the pedipalp into a special pouch in the female, in which the sperm are stored. The female lays the eggs later and sperm are released over them. She then wraps the fertilized eggs in layers of silk to form a protective cocoon, which she hides or carries around until they hatch.

Scorpions

Scorpions (order Scorpionida) have remained virtually unchanged for about 450 million years. They are the largest of the true arachnids, reaching 5 to 8 inches (12 to 20 centimeters) in length. Scorpions are nocturnal and live in tropical and subtropical regions. Their pedipalpi are powerful pincers used to grasp prey. The segmented tail has a sharp sting at the tip with a poison gland, which causes paralysis and death to insects and small mammals.

Other arachnids

Mites and ticks (order Acarina) are small arachnids, usually less than one millimeter in length. Many are parasites, living on blood and tissue fluids, causing skin irritation and occasionally transmitting diseases to their hosts. Their chelicerae can pierce skin, making ticks, for example, difficult to dislodge.

Harvestmen (order Phalangida) are spider-like animals with long legs and only two eyes. Pseudoscorpions (order Chelonethida), like true scorpions, have pincers, but no tail or sting. Sun spiders (order Solufugae), which can be up to 2 inches (5 centimeters) long, have simple elongated pedipalpi that make them look like ten-legged spiders.

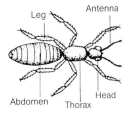

Insects have one pair of antennae and three pairs of legs. The body consists of three parts: head, thorax, and segmented abdomen. Most insects also have wings at some stage in their lives.

The locust has the anatomy of a typical winged insect, with a head, thorax, and abdomen. The thorax has three segments, to which the three pairs of legs and two pairs of wings are attached. A chambered heart forms part of the dorsal blood vessel, which pumps blood forward with the aid of valves. Food is taken through the mouth and is digested in the crop, the pyloric ceca, the midgut, and the hindgut. The first ganglion of the ventral nerve cord forms the brain. Spiracles (not shown) open out in the cuticle of each segment. Air passes into them and carbon dioxide is released from them. The air is conducted by tracheae (ducts) to the blood, into which the oxygen diffuses.

Insects

No other group of land-dwelling animals has as many members as the class Insecta. About one million species of insects have been described—about 300,000 of these are beetles (order Coleoptera)—and some experts believe there may be more than 10 million different species. Insects are the earth's dominant animals. The success of this group is due partly to its tremendous adaptability and huge variation in life styles. Insects live in almost every habitat, from steamy tropical jungles to cold polar regions. The ability to fly has allowed them to spread to new, unexploited habitats and to escape from predators. Many millions of years ago, it helped them to disperse and has given them greater access to food and more desirable environments.

It is possible that the great increase in the numbers and variety of flowering plants during the Cretaceous period also contributed to the enormous success of insects. Most flowering plants are dependent upon insects for pollination.

General features

The exoskeleton, which covers the entire insect body, is composed of chitin and hardened by proteins. It is made up of several parts: the tergum, covering the back; the sternum, on the underside; and two pleura, which link the tergum to the sternum. The pleura are considerably thinner than the rest of the skeleton.

The body is clearly divided into a head, thorax, and abdomen. The head consists of five or six segments, but they are fused together and are not obvious in the adults. Typically, the insect head bears a single pair of sensory antennae, one pair of compound eyes, which may be color sensitive, and one ocelli, or clusters of light-sensitive cells.

A characteristic unique to insects is that the thorax is composed of three segments, each bearing a pair of walking legs. The legs are modified in different species for grasping, swimming, jumping, or digging. The winged insects (subclass Pterygota) also bear a pair of

wings on the upper surface of each of the second and third thoracic segments. The abdomen is made up of 10 or 11 segments connected by flexible membranes. The eighth and ninth segments—also the tenth in males—bear the genital appendages.

An insect's heart lies in the upper part of the body cavity within the thorax and, in most species, in the first nine abdominal segments. The blood circulates in a blood cavity, called the hemocoele, and bathes all the tissues.

Respiration in most insects takes place by means of a system of internal tubes called tracheae, which open to the exterior via paired openings called spiracles. Oxygen diffusion along the trachea is sufficient to meet the demands of the insect at rest. During activity, however, air is pumped in and out of the trachea system by the expansion and collapse of air sacs (enlarged parts of the tracheae). The expansion and collapse of the trachea are controlled by movements of the body. The spiracles can close to prevent water loss.

The principal excretory organs of insects are the Malpighian tubes, which open into the hindgut and the rectum. Uric acid, dissolved salts, and water are drawn from the hemocoele; the fluid in the tubes then passes into the rectum, where useful salts and water are extracted before the waste products are excreted with the feces.

The insect nervous system is much like that of other arthropods, with a brain and a system of linked individual and fused nerve bunches called ganglia connected to a ventral nerve cord. Apart from the eyes, sense organs occur all over the body, but most are concentrated on the appendages. Such sense organs include receptors sensitive to chemical signals and hairs that are sensitive to touch.

Insect flight

One of the most successful adaptive features of insects is flight. Most pterygotes have wings, although wingless insects whose ancestors had wings occur in this group. Various species, such as ants (order Hymenoptera) and termites (order Isoptera), have wings only during certain stages of their life cycle; others, such as fleas (order Siphonaptera), have com-

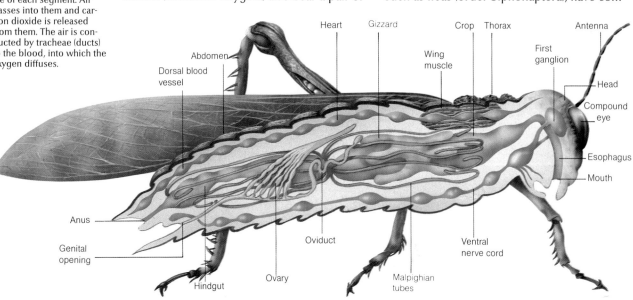

Dragonflies (A) have densely veined, primitive wings that cannot be folded across the back. But more highly evolved insects, such as wasps (B) and beetles (C), have developed structures at the wing base which allow the wings to be folded across their body, and the wing venation is reduced. The two wings on either side of the wasp's body are hooked by frenal hooks. In beetles, the front pair of wings have hardened and become elytra, to form a wing case. They fly only with the hind wings.

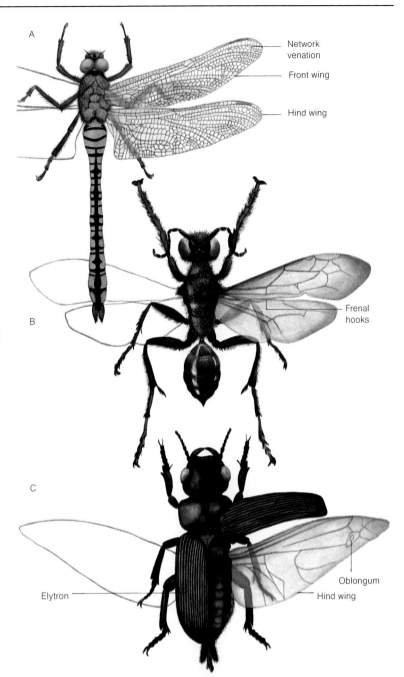

pletely lost their wings as a result of their parasitic life style.

The earliest insect wing is thought to have been a fan-shaped membranous structure with heavy supporting veins. Modern wings, however, are more highly specialized structures, composed of two pieces of opposing cuticle, which are separated in places by veins. These veins support the wing and provide it with blood. Primitive wings such as those of dragonflies (order Odonata) have large numbers of veins that form a netlike pattern, but subsequent evolution has favored a reduction in venation.

Many insects have two pairs of wings. These may move independently, as in damsel flies (order Odonata), or they can be hooked together so that they move as a single structure, as in many hymenopterans, such as bees and wasps. Coleopterans (beetles) have undergone a further change—the first pair of wings has been hardened to form the wing cases, called elytra. These wings form leathery covers that protect the beetle's body. Insects of the order Diptera, which includes all the true flies, have a pair of forewings only. The hind pair has been reduced and modified into club-shaped structures called halteres, which act as organs of balance.

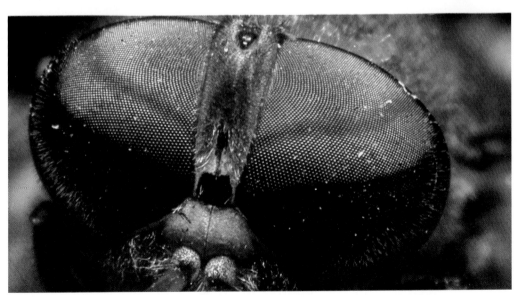

The large compound eyes of flying insects provide good eyesight for feeding. The eyes of some flies each consist of about 4,000 lenses packed together, which are not always of equal size—those of the horsefly are larger on the front and upper parts of the eye.

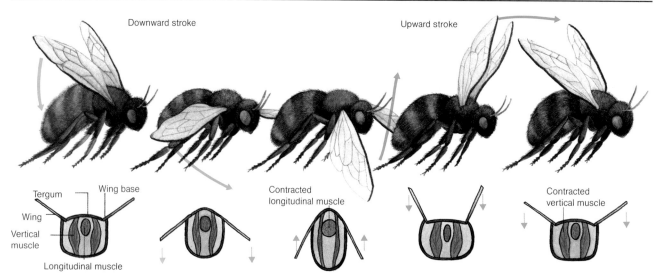

Downward stroke

Upward stroke

Tergum

Wing base

Wing

Vertical muscle

Longitudinal muscle

Contracted longitudinal muscle

Contracted vertical muscle

Insect flight involves a wing movement in the shape of a figure eight. The downward beat results from the contraction of muscles at the wing base and the longitudinal muscles, which also raises the tergum. An upward beat is achieved by the contraction of the vertical muscles.

A honey bee colony consists of workers, drones, and a queen, each with distinct roles. The queen controls the colony and lays up to 2,000 eggs a day. Drones mate with the queen, and workers forage, build combs, and attend to all other aspects of the colony's life.

The action of flight is fairly straightforward. An upward beat is produced by the contraction of vertical muscles in the thorax. When these muscles contract, the thorax flattens, causing the wings to move up. A downward stroke is produced either by the contraction of muscles attached to the wing base or by the contraction of horizontal muscles in the thorax. The thorax arches upward, causing the wings to move down. This up-and-down movement on its own does not provide enough impetus for flight, however, and the insect's wings must also move back and forth, resulting in a wingbeat that forms a figure eight or an ellipse, which is tilted at an angle to the vertical.

The number of wingbeats per second varies greatly from between 4 and 20 for butterflies (order Lepidoptera) to 190 beats per second in bees, and up to 1,000 per second in a gnat. The mode of flying also varies in different groups. Butterflies tend to have a slow, fluttering style, whereas bees and flies can hover and dart. Flight is controlled by a complex interaction of feedback from sensory hairs on the head, stretch receptors at the base of the wing, visual cues, and the wing muscles themselves. There is no nervous center for the control of flight. Flight speed appears to be controlled by the flow of air against the antennae or by sight.

Life cycles and development

Most insects lay eggs. Once hatched, the primitive wingless insects, such as silverfish, gradually mature through a series of molts. They do not change form, but simply grow larger with every molt. The winged insects, however, undergo a metamorphosis in which they change form. These insects are divided into two groups: the Endopterygota, such as flies and butterflies, whose larvae do not resemble the adult and whose wing development is internal, appearing only in the final stage of metamorphosis, and the Exopterygota, such as cockroaches, grasshoppers, and some bugs. These insects hatch as miniature versions of adults, with external wing buds but without mature sexual organs, and they gradually develop by incomplete metamorphosis (hemimetamorphosis).

The basic life style of insects can be modified by various factors. Aphids, for example, exhibit parthenogenesis (reproduction by females without fertilization by males) in favorable weather conditions. Unfertilized eggs laid in the autumn hatch into wingless ovoviviparous females, which do not lay eggs but give birth to broods of similar females. The cycle continues until conditions become less favorable, when one generation produces males along with winged females, and they mate.

Many insects have complex life cycles—especially parasitic insects. Parasitic wasps, for example, have a form of reproduction called polyembryony, in which the embryonic cells give rise to more than one embryo. This means that from one egg deposited in the body of a host, many larvae can be formed, and the resulting embryos can use the host's body as both a refuge and a food source.

A great variety of parasitic insects exist, from blood-suckers that live permanently on a host during their adult lives to the parasites that visit a host only to feed. Some, such as mud daubers, are parasitic in the larval stage

Description	Number in average colony	Average lifespan
Worker Non-reproductive female	60,000 (approximately)	2 — 6 months
Drone Reproductive male	200 (approximately)	2 — 4 months
Queen Reproductive female	1	up to about 6 years

only. The egg-laden female wasp finds a spider, which it paralyzes with its sting. It then builds a nest of mud into which it puts the spider. Finally, it lays an egg on the spider's body and seals up the nest. When the egg hatches, the larva has a ready source of food until it transforms into a pupa.

Feeding adaptations

The mouthparts of insects show great variability. In some, they are adapted for sucking, in others, for piercing and biting. In still others, the mouthparts are modified for chewing, crushing, and tearing.

Chewing mouthparts are found in carnivorous and herbivorous insects. This is regarded as a primitive condition, and is characteristic of insects such as dragonflies, grasshoppers, and beetles, in both the adult and larval stages. Butterflies have chewing mouthparts during the larval stage only.

The mouthparts of bees and wasps are modified for biting and sucking, as well as for chewing. In the case of honeybees, sucking is necessary for the collection of nectar, while chewing ability is necessary for the manipulation of pollen and wax.

The piercing mouthparts of insects that feed on plant juices, such as aphids, penetrate plant tissues, and the insect feeds on the sap. The mouthparts of other insects, such as houseflies, are modified to form a proboscis, which is used to soak up liquid food partially digested by enzymes in the saliva.

Social organization

Communal living is developed to the highest degree in termites, ants, bees, and some wasps. The social life of honey bees, for example, is controlled by a rigid division of labor, and different ranks, or castes, of bees perform different functions within the colony.

A typical colony consists of about 60,000 workers, which are all sterile females; 200 male drones; and a single queen, whose func-

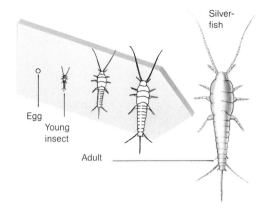
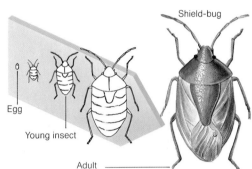
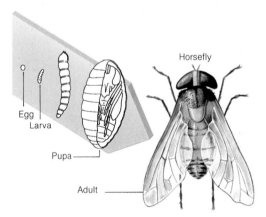
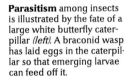

Ametabolous insects, such as the silverfish, develop from the egg stage into a small version of the adult. They develop without metamorphosing, in contrast to hemimetabolous insects, such as the shield-bug, which develops by incomplete metamorphosis into the adult. This insect emerges from an egg, which is shaped like a barrel with a lid. The young insects resemble the adult but are wingless up to the sixth and final molt, when the beginnings of the wings develop. The horsefly is an example of complete, or holometabolous, metamorphosis. Its development occurs in three stages. The first is the feeding stage, when the hatched larva eats continually and increases in size. When it stops feeding, it becomes inactive and constructs a hardened cuticle, or pupa. Inside the pupal case, adult structures develop from embryonic forms. When the period of pupation is over, the adult insect emerges.

Parasitism among insects is illustrated by the fate of a large white butterfly caterpillar (left). A braconid wasp has laid eggs in the caterpillar so that emerging larvae can feed off it.

The peacock butterfly (right) uses both camouflage and warning coloration for protection. When its wings are closed, their dull undersides are good camouflage. But when disturbed, its wings open to reveal prominent "eye" markings, which frighten predators.

Insect mouthparts

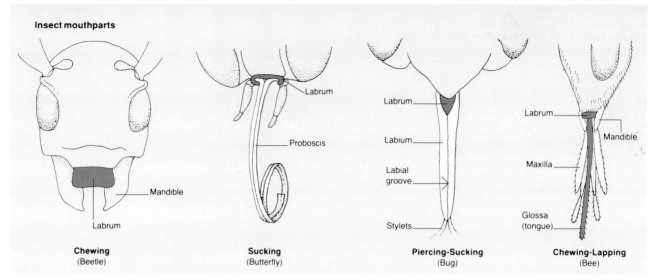

| Chewing | Sucking | Piercing-Sucking | Chewing-Lapping |
| (Beetle) | (Butterfly) | (Bug) | (Bee) |

Insects' mouthparts differ, according to how the animal feeds. Some mouthparts are adapted for sucking, others for piercing and biting, and still others for chewing, crushing, and tearing. Although the two main types of mouthparts are those that are adapted for chewing and sucking, each order of insects has its own variation of one of these types, or a combination of both.

tion is to lay eggs. The main purpose of the drones is to mate with the queen. The duties of the worker bees alter as they become older. Newly emerged workers stay in the hive and clean and feed the queen, drones, and larvae. At this stage glands in the workers' throats produce "royal jelly," which is fed to very young larvae for the first four days of their lives and continually to larvae destined to become queens. After about two weeks, when the wax glands have developed, the workers' duties change to cell construction and cleaning, as well as receiving stores of pollen and nectar from foraging bees. Three weeks later, the worker becomes a field bee, with large pollen baskets on the hind legs. Workers born in the spring live for about eight weeks, but those emerging in late summer or autumn live through the winter. Workers kill or drive drones from the hive in the autumn to conserve food stocks.

The queen maintains unity in the colony by the production of chemicals called phero-

mones that influence the behavior of the bees, largely by inducing the workers to nurse the larvae. As the queen ages, her pheromone production diminishes, and the workers start producing new queens by feeding certain larvae solely on royal jelly. At some stage in the colony's life, the old queen leaves with about half the workers. At this time, a new queen emerges and the other developing queens are usually killed. The new queen mates with the drones during a nuptial flight and returns to head the colony.

A bee hive is made up of a series of vertical wax combs, with hexagonal cells on both sides. Bee larvae mature in the cells. Drone cells are larger than those in which workers develop, and the queen cells are more saclike. Some cells are constructed for storing pollen and honey for use during bad weather or during the winter.

One of the most astounding features of bees is their ability to communicate the position of a rich food source to other workers by means

Termite colonies are controlled by pheromones, which are secreted by the queen. She lies in the colony, trapped by the size of her white abdomen—5 inches (12 centimeters) long—from which she releases several thousand eggs a day. The eggs are fertilized by the only sexually active male (visible in the foreground), who is larger than the other termites. Except for her mate, the queen produces all the members of her colony, which are blind, wingless, and sterile. But at certain times of the year, she produces winged, fertile offspring, which leave the colony.

of a special dance. On returning to the hive, a foraging bee alights on a flat surface and begins to dance in relation to the position of the sun. She moves in a straight line, wagging her body. If she moves upward, the food source lies toward the sun; downward means that it lies away from it. By moving at angles to the left or right, she indicates that the bees must fly with the sun on their left or right, respectively. The speed of the dance indicates the distance of the food. The other workers follow the progress of the dance by touching her, and then fly off toward the food.

Ant colonies are similar to those of bees; there is usually one queen and many female workers, with only a few males. But many ant workers are specialized to perform certain tasks only. Soldier ants, for example, have huge heads and enlarged mouthparts, and their function is to defend the colony. Other ants are workers, tending the larvae and collecting food.

The types of ant colony vary from species to species. They may be groups of hunters, food gatherers, farmers, or even stockbreeders. Army ants are hunters and swarm over victims in enormous numbers, killing them for food. Food gatherers collect plant material and bring it into the nest. So-called "dairying" ants live chiefly on a sugary liquid called honeydew, which they "milk" from aphids and other plant lice. Many ants are fungus growers. Inside their colonies they provide suitable conditions for the growth of particular species of fungi, upon which they feed. Leaf-cutter ants, for example, bring leaf material to the nest on which fungus grows from spores. Other ants keep aphids, which they bring into the nest in winter and put out again in the spring onto suitable plants.

Some species of ants, called slave makers, take ants of other species, which become slave-workers for the colony. Control of the slaves and the colony is generally achieved by pheromones. Ants also have been known to leave their home nest to raid another, which they move into, keeping its original inhabitants captive as slave-workers.

Economic importance of insects

Many insects are of benefit to humans. The caterpillars of the silk moth, for example, provide silk. Farmers raise these caterpillars in huge numbers, feeding them on mulberry leaves. When it is time for the caterpillars to turn into a pupa, they begin to spin a silken cocoon. To extract the thread, the silk farmer must kill the caterpillar and unwind the cocoon. Each cocoon yields about half a mile of fine silk fiber.

Another beneficial insect is the honey bee, which is often domesticated and kept in specially constructed hives, from which honey and beeswax are harvested.

Fruit trees, shrubs, and flowering plants depend upon insects for pollination. Many other insects are useful because they control biological pests. Many ladybugs (family Coccinellidae), for example, feed on aphids, which can severely damage both ornamental and food plants, particularly citrus fruits.

Unfortunately, there are some insects that are pests. Swarms of locusts can lay bare vast

fields of crops, and the boll weevil attacks cotton plants. Some insects are dangerous because they carry disease. Malaria, for example, is transmitted by *Anopheles* mosquitoes. Other diseases transmitted by insects include yellow fever, elephantiasis, sleeping sickness, Lyme disease, plague, Rocky Mountain spotted fever, and typhus. The cost of drugs, vaccines, and eradication programs to control these insects and the diseases they cause is enormous.

Ants milk aphids for their sweet fluid called honeydew, which the aphids collect from plants and release through the anus. But this relationship is not one-sided—the ants build shelters for the aphids and protect their larvae. The ants drink large amounts of honeydew, then return to the nest and regurgitate the honeydew to the ant larvae or other ants.

Molting—called ecdysis—is an essential process for growing insects because their exoskeleton is rigid and does not permit much growth. The intervals between ecdyses are called instars. The mantid, or praying mantis, has from 3 to 12 instars and takes about a year for complete metamorphosis. The insect ruptures the cuticle by swallowing air or water and contracting muscles, which increases pressure inside the exoskeleton. It then wriggles out of the old skin (visible at the bottom of the photograph).

Vertebrates

Despite its appearance, amphioxus is not a fish; it is a lancelet, a chordate animal whose shape reflects the basic vertebrate structure. It has a notochord and segmented muscles.

Animals with backbones, called vertebrates, form only a tiny fraction of all known creatures. In many environments, however, they exert an influence out of all proportion to their numbers, because they are in general larger than invertebrates and in many instances more mobile. Both size and mobility are made possible by a strong internal skeleton, jointed to give flexibility and held in place by tough connective tissue called ligaments.

Structure and development

Vertebrates are part of a larger group, the Chordates (Phylum Chordata). Chordata have a stiff, jellylike rod, called a notochord, running longitudinally through the back part of the body, and acting as an internal support to a series of muscles or muscle segments called myotomes. Above the notochord is a hollow nerve tube, which is usually folded at the front end to form the brain. Below the notochord lies the digestive tract. At some time in their lives all chordates have paired gill slits and a tail. Small, fishlike marine animals called lancelets (subphylum Cephalochordata) have a notochord throughout their life cycle.

In true vertebrates the notochord, present only in the embryonic stage, is replaced by cartilage or bone. Bone is a strong, hard substance, formed mainly of calcium phosphate plus collagen and other protein fibers. Bone has great strength for its weight and is well suited to act as the internal support for the body. It is possible that it originated as a waste product deposited in the skin of prehistoric fishlike marine creatures. When these animals began to live in brackish and fresh water, they faced a change in the balance of their mineral environment. This environment favored the production of bone, because the materials of which it was composed could be used by the body if necessary. This role as a reservoir of certain minerals is still an important function of bone. The early freshwater fishlike vertebrates developed abundant external bone—an enormous store of mineral wealth—which served the additional purpose of protective armor. Such a bony shell is obvious in some modern vertebrates, such as tortoises and armadillos. Though less evident, it also exists in all land vertebrates, in which the large and delicate brain is protected by the bony box of the skull.

The only direct descendants of the original bone-clad vertebrates still surviving are the lampreys (Petromyzonidae) and the hagfishes (Myxinidae), though these have lost all trace of hard skeletal structures. Lampreys and hagfish have a persistent notochord and a gristlelike skull. True fishes, however, have well developed skeletons and, in most cases, an armor of scales made of fine slips of bone set in the skin. The sharks and their relatives have retained bone-based external armor, but also have an internal skeleton formed of cartilage. In other vertebrates, skeletal cartilage is mainly a juvenile tissue, replaced by hard bone as the animals grow.

Vertebrate types

There are eight classes of vertebrates: the hagfish, or Myxini; the lampreys, or Cephalaspidomorphi; the sharks and other cartilaginous fish, or Chondrichthyes; the bony fish, or Osteichthyes; the frogs and other amphibians, or Amphibia; the reptiles, or Reptilia; the birds, or Aves; and the mammals, or Mammalia.

Near the end of the Devonian Period or the beginning of the Mississippian Period, about 360 million years ago, some fishes that were stranded in drought conditions struggled toward new pools using large, strong stiltlike fins. Since that time, vertebrate animals began to become less dependent upon a water environment.

The first creatures to move from the water onto the land were amphibians, which had to return to their ancient habitat to lay their eggs. Young amphibians generally resemble small fishes, but the adults live on land.

Next in evolutionary sequence came the

Vertebrates consist of eight classes: the hagfish; the lampreys; bony fish, such as carp and bass; cartilaginous fish, such as sharks; amphibians, such as frogs; reptiles, such as the green turtle *(left);* birds, such as the sacred kingfisher *(opposite top);* and mammals, such as the oryx *(right).*

reptiles. These developed shelled eggs, enclosing their embryos in the watery white of the egg, which served as their own private pool during the embryonic stage, which corresponds in some ways to the tadpole stage of amphibians. The egg also supplies the embryo's nutrition, in the form of the yolk, which is the perfect, complete food needed for growth until the young reptile hatches. Hatched reptiles show no trace of gills, and their skin is hard and waterproof. Their metabolic pattern requires external warmth, however, so they are restricted to climatically favorable parts of the world.

During the Triassic Period, which began about 245 million years ago, mammals evolved from reptile species that are now extinct. Activity is the keynote of their being, and mammalian bodies are generally well adapted for easy movement. Such activity must be fueled by abundant food, and the relatively high-energy metabolism of mammals generally requires that the body is maintained at a steady temperature—which in animals is aided by insulating hair or fur—while regular breathing provides the large amounts of oxygen such a metabolism needs. The young of almost all species of mammals are born at a relatively advanced stage of development. A special maternal organ, the placenta, provides nourishment for the fetus while it grows within its mother's body. After birth, the mother mammal feeds her young on milk from her body, which gives them all the nourishment they need at this stage. In many species, important maternal and social ties are formed and cemented during this period of suckling. Most mammals have relatively large and complex brains that enable the young to learn behavior, particularly from their parents.

During the Jurassic period, which began about 208 million years ago, birds evolved from reptile stock far removed from the ancestors of mammals. Birds have exploited the possibilities of flight more fully than other vertebrates. To achieve this, they have developed the most energy-intensive metabolic system of all. Birds show many modifications of the general vertebrate pattern, the most obvious being the transformation of the forelimbs into wings. Birds' bodies are maintained at a high temperature and are insulated by feathers. Reproduction still depends on reptilelike eggs. Many birds care for their young for a prolonged period, which establishes social bonds comparable to those seen in mammals.

Fishes

Almost three-quarters of Earth's surface is covered by water, in which live about 21,700 known species of fishes. From their earliest jawless ancestors in the Silurian period, which began about 438 million years ago, fishes have diversified to inhabit widely differing aquatic habitats. Sleek, fast-swimming tuna live in the surface waters of the oceans, while the dark abyssal regions are inhabited by lantern fish and other deep-sea forms. Fresh water contains fishes such as trout—able to survive in the violence of mountain torrents—and, at the other extreme, lungfish, which inhabit temporary pools and can withstand several years of drought. Some fishes have even more specialized habitats, such as the lightless caves of the blind cave characin.

Despite their similar basic design, fish forms show considerable variety. One of the smallest fishes—the pygmy goby of the Philippines, which grows to less than half an inch (13 millimeters) long—is also among the smallest vertebrates. The basking shark grows up to 40 feet (12 meters) long and weighs up to four tons (3.6 metric tons), and the whale shark can reach more than 60 feet (18 meters) in length.

Coloration also varies, from the dazzling multi-colored fishes of coral reefs to the colorless cave fishes, and the almost totally transparent glass catfish.

Fishes are vertebrates, with a backbone of either bone or cartilage. Among their distinguishing characteristics is the fact that they have fins, mostly in paired sets, and that their bodies are covered in scales. Most fishes breathe using internal gills, to which blood is pumped by a two-chambered heart—mammalian hearts, by contrast, have four chambers. Fish are poikilothermal animals—that is, they cannot regulate their body temperature, which varies according to the temperature of the water. The blood of tropical fishes, therefore, is not cold at all. Most fishes are oviparous, or egg-laying, though some species are ovoviviparous—that is, the eggs hatch inside the female's body. A few species, such as the Atlantic manta ray, are viviparous, and the developing embryos are nourished by internal secretions from the female before being born alive.

Fishes are divided into two groups: jawed and jawless. The only jawless species are lampreys and hagfish. Jawed fish are further divided into two groups according to the composition of their skeletons: bony fishes (class Osteichthyes) and cartilaginous fishes (class Chondrichthyes).

Most fishes are carnivorous, preying on smaller fishes or other sea creatures. This grouper is feeding on a shoal of dwarf herrings.

A manta ray feeds on plankton, which it funnels into its mouth with the two "horns" that are extensions of its snout. Its huge pectoral fins propel it through the water by flapping like wings. This photograph also shows several remoras attached by their suckers to the manta's large gill openings, where they feed on plankton that the manta leaves behind, and on parasites from the manta's skin.

The streamlined shape and the segmented body muscles of fishes allow them to use sinuous side-to-side movements of the body and tail to swim through water.

Swimming

Because of its higher density and viscosity, water offers more resistance to movement than air. To minimize drag, therefore, the basic fish shape is streamlined. Backward-pointing scales cover the body, and the skin secretes slippery mucus to cut down water resistance further.

Most fish move mainly by using the caudal, or tail, fin. Muscles on either side of the backbone contract alternately, thus causing the body to bend and the tail to flex from side to side. The caudal fin forces the water backward and this propels the fish forward. Other fins control the direction of movement and help the fish stop. Pectoral fins on the fish's side act as elevators, adjusting the vertical pitch of the fish as it moves. Dorsal and anal fins on the upper- and underside of the fish, which are unpaired, help the fish remain upright. The paired pelvic fins on the underside often act like rudders in changing direction.

In some species, the tail has acquired other functions apart from providing propulsion. The sea horse, for instance, uses its tail to cling to weeds; it swims by lateral undulations of its dorsal fin.

Buoyancy

Cartilaginous fishes are denser than water and so tend to sink. This is described as negative buoyancy. To overcome this tendency they use their pectoral fins, the front of their head, and their tail fin to produce lift as they swim. When stationary, however, they still tend to sink.

In contrast, most bony fishes can give themselves the same density as water—that is, they achieve neutral buoyancy—and are therefore able to maintain their position in the water even when not swimming. They do this using their swim bladder, a gas-filled, lunglike sac, which can be inflated or deflated to alter the buoyancy of the fish, allowing it to rise or descend in the water. Predators, such as the pike, exhibit this ability well, remaining stationary as they wait for passing prey, using only small fin movements to maintain their position.

Sense organs

Most fishes living in clear water have good eyesight. The archerfish, for example, is able to aim a jet of water at insects above the surface and even can adjust for the bending of the light rays as they pass from air to water.

The basic vertebrate shape of a fish is apparent even in a flatfish, where the body appears to have turned onto its side. This computer-colored X ray of a plaice shows the fish's head, skeleton, internal organs, and muscles.

In most fish, the olfactory organs (organs of smell) consist of two pouches, one on each side of the snout. Sharks are renowned for their ability to detect blood in the water from as far away as one-third of a mile (.5 kilometer). Many bottom dwellers, such as the catfish, possess whiskerlike barbels around the mouth, which give them the sense of taste and touch.

Two other sensory systems are found exclusively in aquatic animals, and both relate to water's conducting properties. These are the pressure sense provided by the lateral line system, and an electric sense possessed by some species. In a few species, the latter is also associated with an ability to generate an electrical pulse.

The lateral line system can detect pressure changes caused by disturbances in the water around the fish. In experiments, blinded fishes have been trained to locate a moving glass rod by this method. In the wild, it enables a fish to detect moving food items, approaching predators, or other members of a shoal.

Many fishes that live in muddy water or are active at night also have an electric sense. Electric receptors can be identified in sharks, dogfish, catfish, and mormyrids. Of those species that also possess electric organs capable of generating electrical pulses, some, such as the gymnarchid, can detect irregularities in the electric field and thus locate objects in the cloudiest water. The electric eel has the well-known ability to produce much more powerful pulses (up to 500 volts) when alarmed. It uses such electrical pulses to stun prey.

A fish relies on its fins for movement, balance, and position. The dorsal fins on the fish's back help to keep it upright and traveling straight, though some lateral "weaving" inevitably occurs. The pectoral fins at the sides control its horizontal position.

A   Skeleton of a bony fish

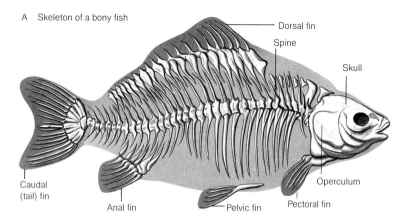

B   Internal and external anatomy

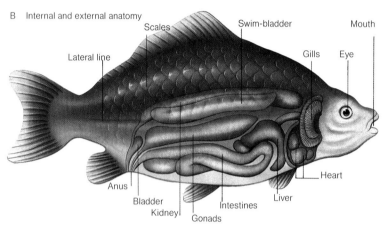

A typical bony fish (A) has unpaired dorsal, anal, and caudal fins—used for balance and propulsion—and paired pectoral and pelvic fins that help control its direction of movement. Internal organs (B) include the heart, liver, and—in advanced species—a swim bladder. The gills (C) absorb oxygen from water that is taken in through the mouth and pumped out through the gill filaments.

C   Horizontal section through mouth chamber

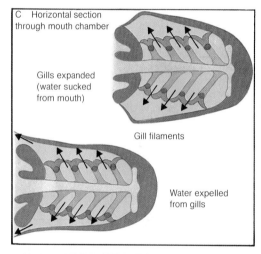

Gills expanded (water sucked from mouth)

Gill filaments

Water expelled from gills

Mudskippers have adapted to an amphibious existence by developing sacs in the mouth and gill cavity that are richly supplied with blood, so that they can absorb oxygen directly from the air.

Gills

The internal gills of fishes have a large surface area over which gaseous exchange between the water and the blood can occur. Gill structure is related to the life style of the fish. Active species need more oxygen and so have large gills. Sluggish forms, like the bottom-dwelling toadfish, have smaller gill areas in proportion to their body size. The mouth and gill cavities form a mechanism that pumps water over the gills. A one-way flow is ensured by flaps of skin in the mouth. The gills are so arranged that the blood in the gill filaments and the water-carrying oxygen flow in opposite directions, and this countercurrent system allows for the most efficient exchange of gases. In some sharks, the respiratory current from the gill slits is strong enough to propel the shark forward.

Certain cells in the gills also help fishes maintain their internal water and salt balance. In freshwater fishes, these cells absorb salts from the water to keep up the body's salt level. In marine bony fishes, other cells excrete salt to compensate for the high levels of salt swallowed in sea water.

Cartilaginous fishes

Because they lack bony skeletons yet possess gill slits, cartilaginous fishes (class Chondrichthyes) were originally considered a primitive group. They do not appear before bony fish in the fossil record, however, and zoologists now believe that their ancestors had bony skeletons. Cartilaginous fishes are divided into three orders: the chimaeras (Chimaeriformes), the rays (Rajiformes), and the sharks (Squaliformes).

About 600 species of cartilaginous fishes have been identified. Most are marine and carnivorous, although the largest representatives are actually harmless filter feeders.

All cartilaginous fishes reproduce by means of internal fertilization. The male cartilaginous fish has a pair of claspers formed from the pelvic fins, which are inserted into the female to transfer the sperm. But after fertilization, cartilaginous fishes show the complete range of different reproductive strategies. Egg-layers lay large, yolky eggs—the familiar "mermaid's purse" is the horny egg case of the dogfish. Offshore sharks are often ovoviviparous—the newly hatched sharks feeding on unfertilized eggs within the female before birth. Some sharks, however, and some rays are viviparous.

There are about 30 species of chimaeras, or ratfish. There are short-, long-, and elephant-nosed species. The upper jaw carries tooth plates and is fixed to the brain case, unlike the mobile jaws of sharks. Chimaeras have long, thin tails and swim by flapping their large pectoral fins. They are medium-sized fish with large eyes, and they live near the ocean bottom.

Sharks, dogfish, skates, and rays all have gill slits and, usually in front of these, another opening called the spiracle. The mouth, on the underside of the head, is filled with rows of teeth, which are developed from the toothlike structures that cover the body. In sharks, only one row of teeth is used at a time, but as soon as teeth are lost or worn away, others move forward to replace them.

Skates and rays (order Rajiformes) have flattened bodies with the gill slits opening on the underside. Some sharks also have flattened bodies, but the gill slits are always on the side of the head. Many skates and rays are bottom dwellers, often lying buried in sand with only the eyes protruding. As with chimaeras, the long thin tail is not used for swimming in most species. Instead they move by flapping or forming ripples in the very large pectoral fins. One of the largest rays is the Atlantic manta, measuring up to 23 feet (7 meters) from the tip of one pectoral fin to the other. Like the largest sharks, it is a filter feeder, straining small fish and crustaceans from the water. Electric rays (suborder Torpedinoidea) use their electric organs—capable of delivering a charge of 200 volts—to stun their prey, or as a means of defense. The stingrays, sluggish bottom dwellers, have a different defensive strategy—a poisonous spine at the base of the tail.

Most sharks and dogfish (order Selachii) have a streamlined, torpedo shape. Probably the most feared and the most likely to be dangerous to humans is the great white shark, which can grow up to 21 feet (6.4 meters) in length. Stomach content analyses of the white shark suggest their normal diet is fish, dolphins, sea lions, and seals. They are also known to attack bathers, but they do not generally eat them.

The smallest members of this group are the dogfishes (family Scyliorhinidae), many of which are smaller than 3 feet (1 meter) in length. Dogfishes live in shallow coastal waters, where they feed on mollusks, worms, and other invertebrates.

Bony fishes

About 95 per cent of living fishes belong to the class Osteichthyes, or bony fishes, making this the largest vertebrate class. As their name suggests, at least part of the skeleton is formed from bone, although cartilage often also occurs. Bony fishes can be readily distinguished from cartilaginous fish by the presence of a bony gill cover called an operculum. Also, the mouth of bony fishes is usually at the

very front of the head, instead of on the underside, as in cartilaginous fishes.

Reproductive strategies also differ from those of cartilaginous fishes. In contrast to the large yolky eggs of the latter, bony fishes tend to lay numerous small eggs. The number of eggs, which can be millions, varies according to the hazards of a particular fish's life style. Fish that lay eggs in areas more vulnerable to predators tend to lay great numbers of eggs. Because male bony fishes do not possess claspers to transfer sperm, fertilization takes place externally.

Another typical feature of bony fishes is their air sac, which functions as a lung in the more primitive forms of fishes, and as a swim

Sharks are voracious carnivores, equipped with a large mouth and rows of sharp teeth to tear flesh from their prey. This photograph also shows the shark's gill arches in the sides of the mouth cavity.

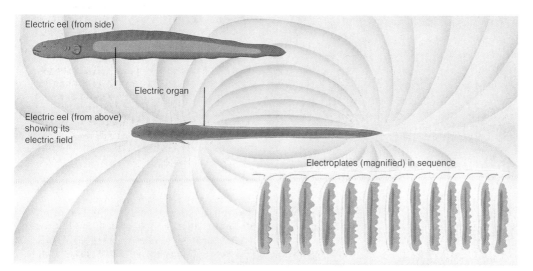

Electric eel (from side)

Electric organ

Electric eel (from above) showing its electric field

Electroplates (magnified) in sequence

Electric fishes can not only sense electric currents in the water but can also generate an electric field, and some species can produce strong electric pulses. The organs responsible consist of rows of electroplates that work neurochemically, intensifying what is essentially a nervous impulse. In the electric eel, these organs are located on either side of the body, and produce an electric field that resembles the magnetic field of a bar magnet.

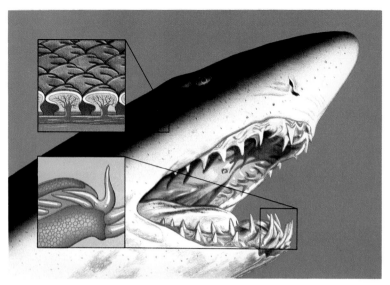

A shark's teeth grow in rows *(lower inset)*. The illustration shows the teeth of a sand tiger shark, which are typical of many shark species. A shark's placoid scales *(upper inset)* have a structure similar to its teeth; this is apparent in the cross section that shows the hard toothlike covering and the soft core of pulp.

oped. Since modern lungfishes have highly specialized jaws with tooth plates, however, they are unlikely to have been the ancestors of four-legged animals. Most zoologists believe that land vertebrates arose from a group of freshwater lobefins. Today, lobefins are represented by a single species, the coelacanth, although more than 50 fossil species are known. This "living fossil" has altered very little since it first appeared in the Devonian period, 400 million years ago.

The lungfishes are also survivors from an earlier period. They were the most numerous fishes in the Devonian period, but there are now only six living species, distributed in Africa, Australia, and South America. All have functional lungs, which allow them to survive in poorly oxygenated waters that occasionally dry up.

Ray-finned fishes

bladder in the rest. Bony fishes are usually subdivided into three groups: lungfishes (subclass Dipnoi), lobefinned fishes (subclass Crossopterygii), and ray-finned fishes (subclass Actinopterygii).

Lungfishes and lobefins

Zoologists have studied lungfishes and lobefins with particular interest since it was realized that similar species probably gave rise to the first land vertebrates. Lungfishes possess choanae (nostrils that connect the mouth cavity to the outside air) as in modern air-breathing amphibians. They and lobefins have paired fins containing bony supports and muscles in their bases, from which the weight-bearing limbs of land vertebrates could have devel-

Most freshwater fishes and commercially exploited marine fishes are ray-finned. As their name suggests, they have fins supported by bony rays. They have two pairs of nostrils, but these do not penetrate to the mouth. Ray-finned fishes fall into four subdivisions: Polypteri, which includes birchirs and reedfish; Chondrostei, which includes sturgeons and paddlefish; and Neopterygii, which includes all other fish. The first two are considered primitive forms.

Birchirs and reedfishes occur in African inland waters. They possess a pair of lungs, so like the lungfishes they can survive in oxygen-deficient swamps. Sturgeons include the world's largest freshwater fish, the beluga, which can grow to a length of 14 feet (4 meters) and a weight of over 2,800 pounds (1,300 kilograms). Sturgeons are also the sought-after

Camouflage is the only means of defense for some fishes. The mottled skin and the flat shape of a plaice enable it to blend with the rocks and gravel of the seabed. Plaice may also try to bury themselves under the sand for even greater concealment.

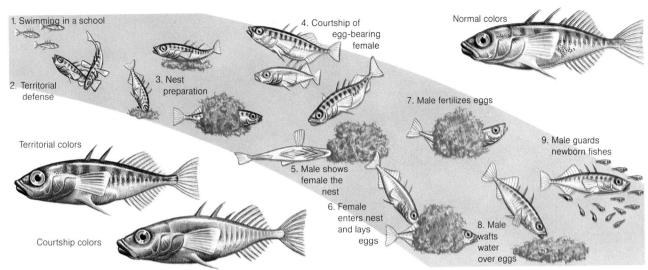

1. Swimming in a school
2. Territorial defense
3. Nest preparation
4. Courtship of egg-bearing female
Normal colors
Territorial colors
Courtship colors
5. Male shows female the nest
6. Female enters nest and lays eggs
7. Male fertilizes eggs
8. Male wafts water over eggs
9. Male guards newborn fishes

source of caviar, which is salted sturgeon roe (eggs). Their skeletons are made mainly from cartilage, and they have several apparently primitive features, such as spiracles and an asymmetrical tail fin. Sturgeons do not have a swim bladder.

The North American gar and bowfin are both primitive neopterygians. They have a lunglike swim bladder and can breathe air when necessary. With sharp teeth, both are voracious carnivores, preying on other fishes and often causing damage to commercial nets in the process.

Teleost fishes

By far the most numerous and diverse group of fishes are the teleosts, which belong to the subdivision Teleostei. Teleost comes from two Greek words meaning *complete* and *bone*. Teleosts are divided into about 30 different orders, of which the most important are eels; salmon, trout, and pike; carps, characins, and catfishes; codfishes; spiny-finned fishes; and flatfishes.

The success of the teleosts can perhaps be attributed to their mobility, achieved by the development of the swim bladder, highly mobile fins, and reduced scales.

The swim bladder allows for buoyancy control, and within the group various stages of swim bladder development can be seen. In the carp, the swim bladder is still connected to the gut by a duct and is filled by swallowing air. In the perch and other advanced teleosts, however, this duct has been lost, and the bladder is filled by gas drawn from the blood at a special "gas gland."

The fins of teleosts also enhance their mobility. The tail fin is symmetrical, giving effective forward thrust. Another feature that can be observed in this group is the tendency of the paired pelvic fins to move forward to underlie the pectoral fins in more "advanced" species. In the salmon, which is considered a less advanced group, the paired fins are well separated, whereas in the more modern perch the pelvic fins have moved forward. Forward pelvic fins appear to aid the mobility of fishes, because stopping and turning are easier with this arrangement.

Other teleost features include a reduction of

the bone in the scales to a very thin layer. Two types of teleost scale occur: cycloid scales, as in the carp, and ctenoid scales, as in the perch. Both types are considerably lighter than the thick ganoid scales of primitive bony fish. Scales are of considerable interest to fisheries biologists in temperate regions, because the age of the fish can be determined by counting the growth rings on its scales.

Eels

Characterized by their snakelike bodies and smooth, slimy, often scaleless skin, eels (order Anguilliformes) are easily recognized. They have no pelvic fins, and their anal and dorsal fins have fused to form a fin seam, which is used in swimming.

Eels have an unusual transparent larval form

In stickleback courtship, the male acquires a red belly as he establishes his territory and prepares the nest. The colors intensify as he identifies a female by her egg-swollen shape, then encourages her to lay her eggs in the nest. Once fertilized, the eggs are tended by the male, who continues to protect the young when they have hatched.

The black marlin is the largest of the billfish, sometimes exceeding 13 feet (4 meters) and 1,500 pounds (700 kilograms).

called a leptocephalus ("leaf-head"), which bears little resemblance to the adult, and was originally considered a completely different fish. Freshwater European eels have an extraordinary reproductive cycle. Adult eels change from freshwater to saltwater fishes, then migrate from Europe to spawning grounds in the Sargasso Sea, an area of the Atlantic Ocean northeast of the West Indies, more than 3,350 miles (5,400 kilometers) away. Once hatched, the larvae make the return journey, aided by the Gulf Stream. Eventually—possibly after as much as three years—they enter European rivers, where they grow into freshwater adult forms.

Most eels are marine, however, like the moray and conger eels, both of which are voracious fish eaters. Garden eels, in contrast, live in tubes and filter tiny zooplankton for food. Despite its name and shape, the electric eel is not a true eel and is more closely related to the carps.

Salmon, trout, and pike

The order Salmoniformes is typified by salmon and trout, both migratory fishes from northern temperate areas, which often use offshore feeding grounds and return to fresh water to spawn. This group also includes the char, grayling, smelt, and whitefish. All can be recognized by their large mouths and the presence of a small, rayless adipose fin between the dorsal and tail fins. All are popular food fishes.

The pike is also a member of this order. A large mouth filled with sharp teeth enables a pike to feed on other fish and even small mammals and birds taken at the water's edge or surface.

Carps, characins, and catfish

The majority of the world's freshwater fishes, more than 3,500 species, belong to the same group as carps and catfishes (order Cypriniformes). All possess a swim bladder connected to the gut, cycloid scales, and well-separated pectoral and pelvic fins.

The 2,000 species of carplike fishes include

most European coarse fishes, such as carp, roach, bream, dace, and tench, plus the non-European goldfish. Carp lack jaw teeth, but possess teeth on the pharyngeal (throat) bones. In the common carp, these crush the plant material and invertebrates on which it feeds. The mouth can also be protruded, which allows carp to suck up food particles effectively. Originally from central Asia, the common carp has been widely distributed and farmed. In 1963, another species, the herbivorous grass carp, was brought to the United States from Asia to help control the overgrowth of certain water plants.

Unlike the carps, characins—which include the brightly colored piranha and tiger fish—have well-developed teeth. Piranha and tiger fish are native to South America and Africa, respectively. Many species of characins, however, like the tetra, have been distributed worldwide to tropical fish enthusiasts, who prize them for their colorful markings.

Catfishes (order Siluriformes), in contrast, are not colorful. They are mainly nocturnal bottom-dwelling fishes. Consequently, their eyes are small, but they bear up to four pairs of conspicuous barbels around the mouth that are sensitive to touch and taste.

All of this group of carps, characins, and catfishes (superorder Ostariophysi) share two interesting features. The first is that they possess a series of three bones (the Weberian ossicles), which connect the swim bladder to the inner ear. The gas-filled bladder picks up and amplifies sound vibrations in the water, which are then transmitted to the ear, giving these fish excellent hearing. To match this, some catfish can also produce sounds by drumming on the swim bladder or scraping the spines of the pectoral fins.

The groups' second common feature is that they all have a similar "fright" reaction. When one fish is injured, its skin cells release a chemical into the water. Other individuals react to this by either fleeing or hiding, so that if a predator attacks one member of a shoal, the others can escape.

Codfishes

Another commercially important group, including the Atlantic cod, haddock, and hake, are known collectively as codfishes or gadids

Moray eels are found in subtropical and tropical seas, where they hide among rocks and in crevices, except when hunting for food. Most morays are aggressive. All have sharp teeth, and some species have glands in the mouth that pour poison into the wound caused by the moray's bite, to disable or kill the prey.

Horse mackerel *(above)* have streamlined bodies and powerful tails. They swim in schools in search of the smaller species on which they prey, and are an important food fish.

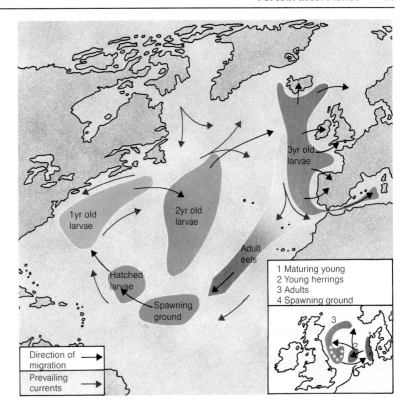

3yr old larvae

1yr old larvae

2yr old larvae

Adult eels

Hatched larvae

Spawning ground

1 Maturing young
2 Young herrings
3 Adults
4 Spawning ground

Direction of migration →

Prevailing currents →

Eels migrate vast distances to spawn *(right),* and their larvae return to European rivers, where they spend their adult lives. Herrings migrate over short distances. The inset map shows North Sea shoals.

(order Gadiformes). They are essentially cool sea fishes, living mainly in the northern waters of the Atlantic and Pacific oceans. Hake, however, also live in deep water, and have invaded subtropical areas. At night, they rise to the surface to feed on other fish, such as herring. Growing up to 1 foot (30 centimeters) in length, hake now form the main catch of fishing industries in Europe, South Africa, and South America.

Spiny-finned fishes

The most advanced teleosts have spiny rays in the dorsal, pelvic, and anal fins. Advanced features of teleosts include ctenoid scales, a swim bladder not connected to the gut, and pelvic and pectoral fins placed close together.

There are more than 7,000 species of perch-like fish of the order Perciformes, which include the angelfish, butterfly fish, mackerel, and tuna. They are active, often brightly colored fishes, with good eyesight and color vision.

The European perch is an active predator, feeding on other fish—as is the much larger Nile perch, but within this group all manner of feeding specializations are found. One family, for example, Cichlidae, or the cichlids, are particularly diverse. In southeastern Africa's Lake Malawi alone, 200 different species of cichlids coexist, feeding according to their separate tastes on mollusks, insects, fishes, eggs, plankton, algae, or even the scales of other fishes. In each case, the pharyngeal teeth appear specially adapted to the diet.

Flatfishes

Many teleosts are flattened from side to side, but this is taken to extremes in the flatfish (order Pleuronectiformes).

The larvae of flatfish are conventional in appearance, but as they grow they gradually become compressed and turn over on one side. One eye begins to move closer to the eye on the opposite side of the head, and the mouth becomes twisted. The underside becomes white, and the upper surface darkens.

This shape is ideal for the flatfish habitat. Most live on the seabed in coastal waters, and some, such as the plaice, possess the ability to alter their coloration to match their surroundings. Commonly eaten flatfish, such as sole, are small, although others, such as halibut, reach weights of 400 pounds (180 kilograms).

The coelacanth, a "living fossil," may be a link between fishes and amphibians. Once thought to be extinct for 60 million years, it was discovered in 1938 in the waters off southern Africa.

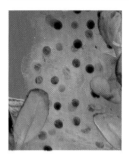

Frogspawn consists of the fertilized eggs of a frog, with the dividing cells surrounded by a gelatinous envelope. The eggs of other egg-laying amphibians such as toads, newts, and salamanders, look similar.

Amphibians

According to paleontologists, the oldest fossils of amphibians date back to the end of the Devonian Period—more than 350 million years ago. The Devonian Period was a time of great ecological change, which resulted from massive disruptions of the surface features of the earth. For the evolution of amphibians, the most significant of these were changes that altered the level of the seas.

As water levels rose and then receded, the sea left behind organic materials in which plants could thrive. This tended to encourage the development of areas of lush vegetation in swamplike coastal regions. These conditions favored creatures that could get their oxygen from air as well as from water, and it is probable that it was in such an environment that the first amphibians evolved.

In prehistoric times, there were at least six orders of amphibians (class Amphibia), but now only three orders remain: Urodela, or newts and salamanders; Anura, or frogs and toads; and Apoda, or caecilians. Although the majority of amphibians inhabit the tropics, representatives of the class are found throughout the world, except in places where there is no water at all. In cold climates, amphibians hibernate during winter. Like most animals except birds and mammals, they are poikilothermal, which means that their body temperature changes with fluctuations in the temperature of their environment.

Amphibians vary greatly in size, from the African sedge, or reed, frogs of the *Hyperolius* genus, which measure an inch or less in length, to the Japanese giant salamander and a caecilian from Colombia at more than 5 feet (1.5 meters) long.

Amphibians have a wide flat skull attached to the spinal column, which may be short, as in frogs and toads, which usually have eight vertebrae in the trunk, or extremely long, with as many as 250 vertebrae in some salamanders and caecilians. The urodeles and anurans have four limbs, with four fingers on each forelimb and five toes on each hindlimb. The apods have lost their limbs. Unlike the urodeles, adult anurans have no tails.

The skin of all amphibians is toughened (cornified) on the animal's upper surface, and smooth on the lower surface. All shed their skin regularly. The skin produces slimy or poisonous secretions that make these animals unpalatable to most predators and so afford protection. In many species, the skin color can change, usually for camouflage or mating purposes. The skin also plays a part in respiration.

Respiration

In evolutionary terms, one of the most profound modifications needed for animals to survive the emergence from an aquatic environment was the ability to absorb oxygen from air rather than from water. In amphibians, perhaps reflecting stages in this transition, respiration takes place in gills, lungs, the lining of the mouth, or the external skin, either alone or in combination, depending on the species and the stage of development. The role of the skin in respiration varies. In some species it is not particularly important, while for others the skin is the sole means of absorbing oxygen. For example, the skin of some lungless salamanders, such as the European alpine salamander, absorbs oxygen directly from the aerated water of the mountain streams in which the salamanders live.

Senses

Amphibians exhibit a variety of sensory needs. In almost all species the sense of touch is well developed, but the development of organs for sensing things at distance—for sight, hearing and smell—varies greatly. Cave dwellers, such as the olm, a type of European salamander, have little need for sight and their eyes are undeveloped. The eyes of burrowing caecilians are either very small or completely absent, although caecilians have another organ similar to a small feeler, which is associated with eye muscles and seems to have a sensory function. In other species, particularly frogs, the eyes are well developed. Underground and cave-dwelling amphibians tend to have a good sense of smell, and this is also important for aquatic species, unless they live in clear water, where good eyesight is more beneficial.

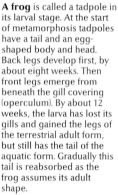

A frog is called a tadpole in its larval stage. At the start of metamorphosis tadpoles have a tail and an egg-shaped body and head. Back legs develop first, by about eight weeks. Then front legs emerge from beneath the gill covering (operculum). By about 12 weeks, the larva has lost its gills and gained the legs of the terrestrial adult form, but still has the tail of the aquatic form. Gradually this tail is reabsorbed as the frog assumes its adult shape.

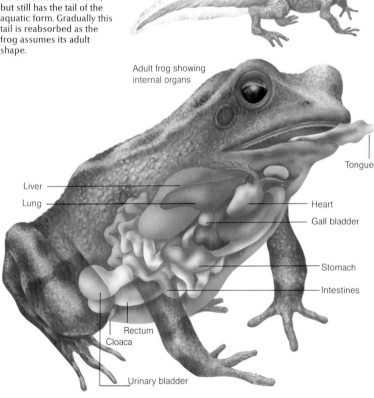

8 weeks

Mouth
Eye
Operculum

12 weeks

Adult frog showing internal organs

Tongue

Liver
Lung
Heart
Gall bladder
Stomach
Intestines
Rectum
Cloaca
Urinary bladder

Lateral line organs are present in aquatic amphibians, although they are less sophisticated than in fishes. They appear as lines of small pits along the head and body, distributed in relation to the lateral line nerves. They respond to vibrations or changes of pressure in the water, enabling the animal to control its equilibrium and posture and to detect predators or prey.

The existence of a sense of hearing in amphibians seems to depend on whether or not the animal has a voice. Urodeles and apods have neither voice nor visible ears, but appear to be able to detect vibrations. Frogs and toads, in contrast, have remarkably loud, even strident voices, and have an eardrum just behind each eye.

Movement

In water, amphibians can both walk and swim. During courtship, for example, newts often walk along the bottom of a pond. Frogs and toads use their hindlimbs for swimming, whereas newts propel themselves with their tails, keeping their hind legs pressed together. Caecilians swim like eels. Although urodeles can walk on land, their tails and short legs restrict their mobility. Frogs and toads crawl on land, but also use their powerful hind legs for leaping—several species can jump 20 times their body length. The limbless caecilians usually burrow below the ground. When observed aboveground, however, their movements resemble those of a snake.

Feeding

Apart from anuran larvae (tadpoles), which eat mostly plant material, amphibians feed chiefly on insects and small invertebrates. Caecilians have a more varied diet, which includes other amphibians, fishes, and even some reptiles, small mammals, and birds.

Amphibian teeth have no roots and grow all the time as they are worn down. Some urodeles have fixed, immobile tongues, but others, such as the cave salamander, have tongues that can protrude to catch prey. This ability is particularly marked in frogs and toads, in which the tongue is attached to the front of the lower jaw. As the tongue flicks forward, it scrapes the roof of the mouth, collecting a sticky covering that adheres to the prey,

Frogs have adapted successfully to many environments. The Costa Rican flying frog is a tree frog with webbed feet that enhance its ability to leap to such an extent that it appears to fly.

so that it can be drawn back to the amphibian's mouth.

From the mouth, food passes to the stomach, where secretions from stomach glands start the digestive process. This process continues through the intestine, aided by enzymes from the pancreas and bile from the gall bladder, to the rectum and cloaca. Urine also empties from the urinary bladder into the cloaca, from which waste products, in the form of urine and feces, are expelled.

Reproduction and life expectancy

Most amphibians lay eggs, and fertilization usually takes place outside the body. The young pass through a larval stage, although in tailed amphibians and caecilians the distinction between the larval and the adult form is less obvious than it is in anurans. Some species (most caecilians, for instance) bear live

The fire salamander protects itself with poison glands in its skin. These secrete an irritant fluid that burns the mouth of any predator that bites the salamander.

Amphibians feed chiefly on insects, and some are agile predators. In this photograph, a palmate newt has captured a fish.

young. Others lay eggs that contain partially developed young, and still others (such as the common European frog) develop from embryo through the larval stage in the egg.

In captivity, amphibians have been known to live for 15 years or more, but their life expectancy in the wild is usually much shorter. Among the longest-lived are some slender salamanders and cave salamanders, which live for as long as 10 years in their natural habitat.

Frogs and toads

The voice of a frog is produced when air is pumped over the vocal cords in the throat. Male frogs of some species—in this photograph the painted reed frog—have a vocal sac that can be distended to produce a resonating chamber, which intensifies the sound considerably.

The anurans, loosely called frogs and toads, form the largest of the three orders of Amphibia, with some 2,700 species. Anurans have adapted to a wider range of habitats than the other orders and live on every continent except Antarctica. Most are terrestrial rather than aquatic, and some, such as the tree frogs (Hylidae), dwell primarily if not exclusively in trees. In some of these species, eggs are laid

in water trapped among leaves, but some build nests and some spawn in ponds and similar locations, then return to their arboreal habitat.

The terms frog and toad are based on appearance and do not relate to actual evolutionary distinctions. Anuran classification is based primarily on skeletal features, such as the presence of ribs, which distinguishes lower from higher anurans, or the structure of the pectoral and pelvic girdles. Beyond such relatively simple criteria anurans show enormous variety. Frogs and toads vary according to the extent to which hands and feet are webbed, in the shape of and function of fingers and toes, and in the distribution or arrangement of teeth. Some species have no teeth, others have teeth in the upper jaw only, and one genus, *Amphignathodon,* has teeth in both jaws.

Most anurans are nocturnal, although newly metamorphosed frogs take several days to assume their nocturnal habits. They are mainly dependent on their eyes for catching prey—their visual system responds to small, irregularly moving objects, and they will not react to even preferred prey unless it moves.

An important feature distinguishing anurans from other amphibians is their voice. Almost all anuran males can increase the volume of the sounds they make by using vocal sacs in the mouth. The mating call of the natterjack toad, for example, can be heard more than half a mile away. Females, which are generally larger than males, have a quieter call.

In the breeding season the male anuran's call serves to attract females and enables a female to recognize a male of the same species. When a male anuran is ready to mate he will jump upon anything of the appropriate size, and will often clasp a male or some other inappropriate object (even a piece of mud or earth) before managing to find a female.

In most frogs and toads, eggs and seminal fluid are emitted at the same time. The female then deposits fertilized eggs singly or in clumps or strings. The majority of European and North American anurans lay clusters of hundreds of eggs, each in its own gelatinous envelope. One exception is the North American tailed frog, which lays eggs that have been fertilized internally. Internal fertilization also occurs in one genus of toads, *Nectophrynoides,* found in Africa, which do not lay eggs at all, but give birth to live young.

Salamanders and newts

The tailed amphibians (Urodela) make up the second largest of the three orders of Amphibia. There are about 330 species of salamanders and newts, most of which inhabit the temperate regions of the Northern Hemisphere. Urodeles live in a wide variety of habitats. Many are terrestrial, retreating to moist crevices in hot, dry weather. Some live in trees, some never leave the water, and some live in the total darkness of deep caves. Like frogs and toads, the tailed amphibians need moisture; insects, worms, or even fishes for food; and some species need fresh water in which to lay eggs.

The mating behavior of salamanders and newts differs from that of anurans. Salamanders and newts have no voice, so mating dis-

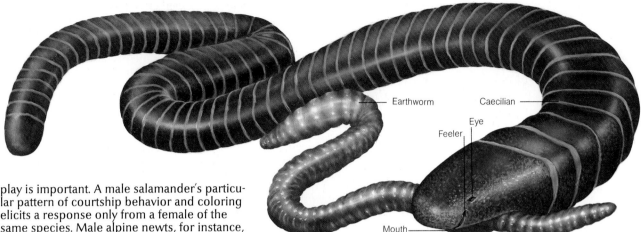

Earthworm   Caecilian

Eye

Feeler

Mouth

play is important. A male salamander's particular pattern of courtship behavior and coloring elicits a response only from a female of the same species. Male alpine newts, for instance, develop crests and striking coloration—blue on the back and flanks, orange on the belly—in the breeding season. Many species indulge in complex courtship rituals, ranging from lashing the tail in order to waft fragrant secretions toward the female to swimming back and forth in front of her.

During mating the male deposits a spermatophore, or sperm packet, which the female takes up into her cloaca. The sperm are then released as the female releases her eggs. When they hatch the larvae resemble adults, except they have gills that disappear during metamorphosis.

Caecilians

The third amphibian order, and the one about which least is known, is the Apoda. Apods, or caecilians, have no limbs and look like large worms or smooth snakes. Lengths vary from just over 2 inches (6 centimeters) to nearly 5

feet (1.5 meters). Nearly all species live in warm temperate and tropical regions. They usually burrow beneath the ground, and are seldom seen above ground in daylight, although it is believed that most come to the surface at night. Eyes are of little use in such a habitat, but caecilians have developed a sensitive "feeler" or tentacle that probably helps them search for the worms and insects that form most of their diet. Reproduction is by internal fertilization, and it is believed that caecilians either lay eggs or retain the eggs until the young hatch.

Caecilians are sometimes mistaken for very large earthworms, but when the two are compared—as in this illustration of a South American caecilian eating an earthworm—there are obvious differences. Most significant of these are the caecilian's flattened head; its small, weak eyes; the small feeler located between its eye and nose; and its mouth. Some caecilians are strikingly colored: the species illustrated is blue-black, darker above than below. Others are mottled or striped.

In courtship, many species of newts develop striking coloration. In some species, the crest of the male also enlarges. Both characteristics are evident in this photograph of great crested newts mating.

Reptiles

Reptiles were the first true land-dwelling vertebrates, appearing about 310 million years ago. They dominated Earth from about 225 to 65 million years ago, but today only four reptilian orders remain: Testudinata (or Chelonia), the turtles; Squamata, the lizards and snakes; Crocodilia, the crocodilians; and Rhynchocephalia, the tuatara.

The most noticeable characteristic of reptiles is their outer covering of scales. They also lay tough, leathery eggs and are poikilothermal—that is, they lack an internal means of controlling body temperature. Because of this last feature, most of the 6,000 or so reptilian species live in the tropics or subtropics. However, some live in temperate regions, including the tuatara and the viviparous common lizard, which is found as far north as the Arctic Circle.

The reptilian body

A reptile's body is covered by a dry, relatively thick waterproof skin, which helps to protect the internal structures and prevents dehydration. The scales on the reptilian body are really folds in the skin, composed of a hard, transparent dead material called keratin. In chelonians and crocodilians, the outer keratin layer is continually worn away but a new layer always remains underneath. Snakes and lizards, however, shed their scales at regular intervals. Lizards often shed their scales in strips. But some, like snakes, shed the whole skin in one piece, a process known as sloughing. Lizards and some crocodilians have pieces of bone called osteoderms within their scales, which give the skin additional toughness.

The skeleton of most vertebrates, including reptiles, consists of bone and cartilage. The amount of cartilage in the body is highest in young reptiles, but as the reptile becomes an adult, the cartilage is usually present only over

The Stegosaurus, a dinosaur that lived about 150 million years ago, may have regulated its body temperature by means of the blood-rich skin that covered the diamond-shaped plates on its back. Some authorities believe that the blood absorbed heat from the sun and carried the heat around the body. In the shade, the blood was cooled.

The reptilian skeleton varies greatly. The crocodilians have abdominal and thoracic ribs; in the chelonians, or turtles, the ribs and vertebrae are fused to their shell. Snakes' ribs extend to the end of the tail.

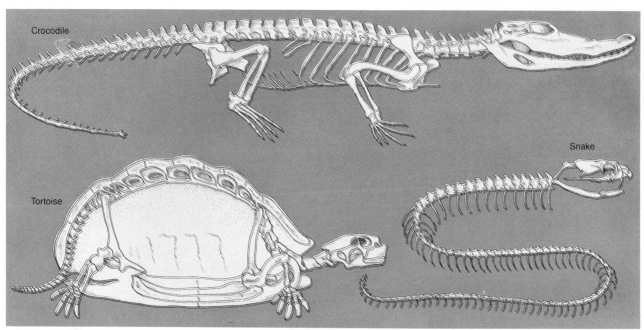

Crocodile

Snake

Tortoise

the joints, the shoulder girdle, and parts of the skull. Unlike mammalian bones, those of a reptile continue to grow throughout the animal's life.

The ribs of reptiles extend the length of the body, from the third vertebra of the spine to the beginning of the tail, whereas in mammals, they are confined to the chest region. The skull consists of many separate bones. Similarly, the lower jaw is formed from several distinct but joined bones. The teeth are peglike or pointed and are all alike. Reptile teeth are not present in a variety of forms as are the teeth of mammals.

The reptilian brain is small in relation to the body—in many, it makes up less than 0.05 per cent of the total body weight—and reptiles are relatively unintelligent, having very little ability to learn or adapt to new situations. But they compensate for this deficiency by possessing elaborate patterns of instinctive behavior that enable them to survive and reproduce successfully.

Reptiles breathe with the aid of two lungs, one or both of which may be well developed. Some species can also absorb oxygen in the water through the membranes that line the mouth and the genital and waste-eliminating chamber called the cloaca. The cloaca is at the end of the large intestine, and opens under the body in front of the tail. It receives eggs or sperm from the sex glands as well as feces from the intestine and urine from the kidneys. The water needed for elimination of waste products is reabsorbed into the blood through the walls of the large intestine and cloacal chamber. As a result, the anal waste is solid or semi-solid, and there is little or no liquid urine. In this way, water is conserved in the body, a feature that is important for those reptiles that live in dry environments.

Senses and sense organs

Most reptiles have good eyesight. The fields of vision of each eye overlap to some extent, so that some of these animals have binocular vision, although sometimes to a limited degree. Vision is particularly highly developed in chameleons, which must be able to judge distances accurately in order to catch their prey.

Relatively little is known about hearing in reptiles, although this faculty does not seem to be as important as vision. Like birds, mammals, and amphibians, reptiles have a small bone (the stapes), which transmits sound vibrations from the eardrum to the inner ear. Most reptiles, except for snakes and a few lizards, have external ears with eardrums on the surface or sunk down on a tube at the back of the head; and, except for crocodiles, they lack earflaps. Snakes also lack the middle ear cavity—they "hear" a sound mainly by picking up vibrations from the ground, which are then transferred from the lower jaw bone to the skull.

Smell is an important sense in most reptiles. Snakes and lizards have two tiny cavities known as the Jacobson's organ, which work in conjunction with the tongue as well as with the sense of smell. These organs are found above, and open into, the roof of the mouth and are partly lined with sensitive membranes similar to those that line the nose. As the

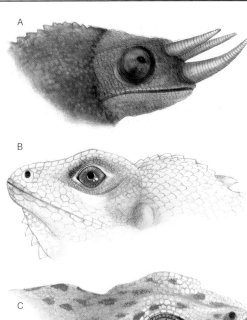

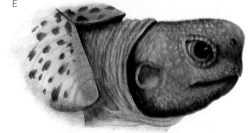

Reptilian eyes are varied. The chameleons' rotating eyes (A) allow them to keep still while watching for prey. Their eyelids cover most of the eye, which sharpens their focus. The round pupils of some lizards (B) contrast with those of the tokay gecko (C), which are vertical and constricted in four places to protect them from dazzle and improve vision. Crocodilian eyes (D) are covered with a membrane that allows them to be submerged in water. Tortoises (E) have round pupils protected by thick, folding eyelids.

Many species of lizard have spines, shields, or frills on their body or head, such as those found on the bearded dragon lizard. These features are either modified scales or skin membranes. They are puffed up to give the lizards an appearance of greater size in displays of aggression or in courtship.

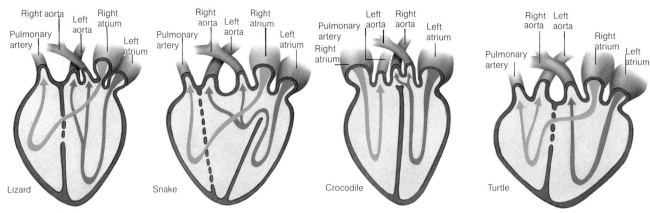

Lizard   Snake   Crocodile   Turtle

The heart system differs among reptiles. In lizards, a perforated septum separates the ventricles. Nonaerated blood flows through it from the left ventricle into the pulmonary artery in the right ventricle. In the snakes the pulmonary artery, the aortas, and the right atrium are in the right ventricle. A muscular ridge formed during contraction guides the nonaerated blood into the pulmonary artery. The septum of the crocodiles is whole, and the two aortas come out of each of the ventricles. This is also true for chelonians but, because their ventricles are not perfectly divided, nonaerated and aerated blood flows into the aortas, so a mixing of blood occurs.

tongue moves in and out, it picks up minute particles from the air or off the ground, which are carried to the Jacobson's organs.

An additional sense organ—the pineal eye—occurs in the tuatara and many lizards. It was also present in many reptiles and amphibians now extinct. It is found at the top of the head and opens, by means of a small hole, into the skull. It is covered by skin, which usually has a lighter color than the surrounding skin. The function of this "third eye" is unknown, but it is light-sensitive and probably helps the animal to regulate its exposure to the sun and thus to control its internal body temperature.

Temperature regulation

The term "cold-blooded" is misleading when applied to reptiles because under certain circumstances, a reptile's body temperature may be very high, even higher than 98.6° F. (37° C), the normal body temperature of humans. Reptiles are, in fact, poikilothermal—they do not have an internal mechanism to regulate body temperature—and therefore depend on the external temperature. Body temperature is important because it is related to activity. Under extreme conditions of heat or cold, the body becomes sluggish, because vital activities

such as heartbeat and breathing slow down. But within a narrow temperature range—which varies from species to species—a reptile reaches its highest level of activity. For most, the optimum temperature is between 77° F. (25° C) and 99° F. (31° C).

Reproduction

Most reptiles that live in temperate and subtropical climates breed in spring. As in most other animals, the urge to mate is triggered by an increase in the hormone levels in the sex glands and other internal organs. These changes are usually a response to changes in the environment, such as lengthening of the day, an increasing abundance of food, and a rise in air temperature. Tropical reptiles may breed several times a year, but even in these species mating usually follows a seasonal cycle.

With the exception of the tuatara, which mates by pressing its cloaca against that of the female, all male reptiles possess a penis through which sperm is introduced into the female's cloaca. Chelonians and crocodiles have a single penis, whereas lizards and snakes have two, known as hemipenes, only one of which is used during copulation.

Basking on warm rocks prevents such reptiles as marine iguanas from becoming too cold. They move into shade periodically to keep from overheating. This dependence on external temperature is essential, because they have no internal mechanism to regulate their body temperature. In addition, some reptiles change the color of their skin to make it more or less heat absorbent.

Most reptiles are oviparous—that is, the young hatch after the eggs have been laid. But some species of snakes and lizards are ovoviviparous (the young hatch while the eggs are still inside the mother's body). In such reptiles, the young may still be surrounded by embryonic membranes when they are born. Exceptions to both these methods of reproduction are found in certain skinks and snakes, in which the relationship between mother and embryo is more intimate, resembling that of mammals. In these reptiles, a type of placenta develops that lies close to the lining of the mother's oviduct.

Tortoises, turtles, and terrapins

Members of the order Chelonia—tortoises, turtles, and terrapins—differ from one another in the following way: tortoises live only on land, turtles live in the sea, and terrapins live in fresh water. In the United States, however, most chelonians are commonly referred to as turtles.

The most striking feature of this order is the presence of a shell. The shell is composed of an outer layer of horny structures called scutes, which are formed from skin tissue, and a thick inner layer of bony plates. The outer, horny layer increases in size with age, but its growth pauses during hibernation. Each dormant phase produces a ring around the previous year's growth, similar to a growth ring in a tree. Soft-shelled turtles and the leatherback turtle have an outer layer of tough skin rather than scutes. The shell of the hawksbill turtle is the source of the commercial material known as tortoiseshell.

Chelonians do not have teeth. Instead they have a horny beak on the upper and lower jaws. Tortoises are mainly herbivorous, turtles eat a mixture of plant and animal food, and terrapins tend to be carnivorous, feeding on invertebrates and fish.

There are two suborders of chelonians, distinguished by the way the head is withdrawn into the shell. The Pleurodira, the side-necked turtles of South America, Africa, and Australia, are a small group of two families. An example is the South American matamata terrapin, which has a long snout with nostrils at the end and a long flap of skin behind each eye. It feeds on fish and other small aquatic animals.

The Cryptodira, or hidden-necks—a group that comprises seven families—can bend the neck vertically up and down. The family Testudinidae in this group includes the land tortoises, which are some of the most familiar chelonians, with about 40 species. The family Emydidae, with about 80 species, includes terrapins and pond tortoises. The pond tortoise and the red-eared terrapin are commonly kept as pets.

The family Chelydridae is found in North America, Central America, and northern South America, and is comprised of the musk turtles, the mud turtles, and the snappers. The common snapper is a ferocious freshwater turtle that bites sharply when disturbed. Unusually, it cannot withdraw its head into its shell, and its tail, legs, and neck are almost completely exposed on the underside. The musk turtles are exceptional among reptiles in having scent glands on their body. They are often called

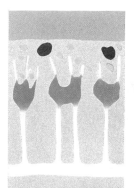 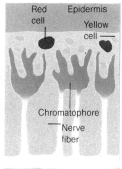

Red cell   Epidermis   Yellow cell

Chromatophore
Nerve fiber

Color change in chameleons is due to pigment-containing cells under the skin's surface. When the pigments of these branched cells are concentrated at the cell center (far left), the skin is light colored. When the pigments spread into the cell branches (right), the skin darkens.

"stinkpots" because of the musky odor they emit when handled.

The sea turtles (families Cheloniidae and Dermochelyidae) are the most highly adapted to a watery existence. Their legs are flattened and paddle-shaped and they swim with agility. On land, however, they move clumsily and slowly. Usually only the females come ashore, specifically to lay eggs in a hole in the sand.

Most reptiles lay their eggs on land, usually concealed under a stone or buried in soil or sand. Some, such as rat snakes, lay them in tree hollows. The eggs are protected by a relatively tough, parchmentlike shell.

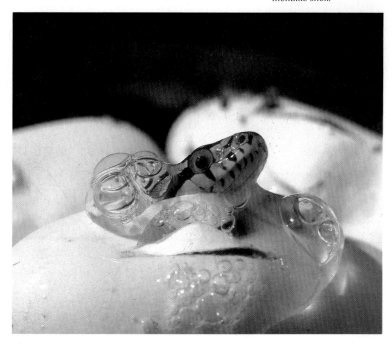

Lizards and snakes

These reptiles belong to the order Squamata, which is divided into two suborders: the Sauria, or lizards, and the Serpentes (also called Ophidia), or snakes.

Lizards and snakes form the largest group of reptiles and are most abundant in tropical regions. Most lizards are small and four-legged with long tails, movable eyelids, and external ear openings. Many can shed their tails to escape from predators. The tail is then regenerated but is never an exact copy of the original, being shorter and with an irregular scale pattern. Members of the family Anguidae have either one or two pairs of short, reduced limbs and some, such as the slowworm, are legless.

An interesting feature of some skinks (family Scincidae) is that the eyelids contain a transparent opening, which allows them to see while they close their eyes to protect them against flying debris.

The family Chamaeleonidae contains some 100 species of chameleons. These lizards are exceptionally well adapted to an insectivorous and arboreal life. Two of the five toes on each limb are opposed to the other three, forming a gripping "hand," and the tail is prehensile. Each eye can move independently or both can focus on the same object. The chameleon can therefore judge distances accurately and aim precisely at the insects on which it feeds, catching them with its long, sticky tongue.

A lizard called the anole is closely related to the chameleon. There are more than 225 species of anoles. They are often called American chameleons.

Only two species of lizards are venomous: the Gila monster and the beaded lizard, both of which live in desert areas of North America. The venom is secreted along the grooves in the teeth at the base of the animal's lower jaw.

The snakes, however, contain most of the poisonous reptilian species, although the majority of species are harmless. Venomous species are classed in two main groups—vipers and elapids—according to the position and construction of the fangs. Vipers include rattlesnakes, the adder, and the asp. When their mouth is opened, the fangs swing down and forward. Elapids include the cobras, mambas, and kraits, in which the front fangs are fixed. Most venomous snakes are front-fanged, but some are rear-fanged. The snakes in the family Boidae, such as the boa constrictor and the pythons, kill their prey by crushing it.

Unlike lizards, snakes are limbless, although some, such as the boas, have rudimentary hind legs. The tail cannot be regenerated when lost, and the transparent eyelids are fused together to form a protective covering. In addition, most snakes are able to dislocate their jaws, enabling them to swallow prey considerably larger than themselves.

Crocodilians

All the members of the order Crocodilia—crocodiles, alligators, caymans, and gavials—are large, amphibious reptiles. Crocodilians use their long, vertically flattened tail for swimming. They usually keep their webbed feet tucked in at the sides of their body. The ears, eyes, and nostrils are located high on the head, so that the animal can remain hidden under the water and at the same time can hear, see, and breathe. When a crocodilian is totally submerged, the eyes are covered by a membrane, and the ears and nostrils are closed by special valves. In addition, a fold of skin shuts off the windpipe so that the animal can open its mouth underwater without drowning.

Crocodilians are carnivorous, and adults prey on large animals—occasionally as large as a horse—which they catch either underwater or on land. Crocodilians eat their prey whole or first tear it apart and the eat it. They

The giant tortoises are found in the Galapagos Islands, Aldabra Islands, and the Seychelles Islands. They may measure up to 4 feet (1.2 meters) long, and they are known to live to about 100 years of age. They live in the lowlands but venture to the highlands to find pools for drinking and wallowing.

The venomous snakes are grouped into three families according to the position of their fangs. The Elapidae, including the cobras and mambas, have fangs fixed to the front of the mouth. Viperidae also includes front-fanged snakes, such as vipers and adders, but these snakes have long fangs on a rotating maxillary plate that lie against the palate when the mouth is closed. As they open their mouth, the fangs swing down and forward. Some venomous snakes, such as rat snakes, have one to three grooved fangs on the upper jaw in the rear of the mouth. The poison runs down the groove in the teeth and is injected into the victim when the snake bites it.

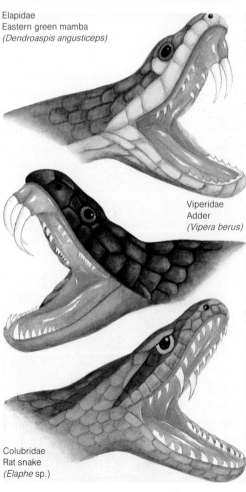

Elapidae
Eastern green mamba
(*Dendroaspis angusticeps*)

Viperidae
Adder
(*Vipera berus*)

Colubridae
Rat snake
(*Elaphe* sp.)

often store large prey underwater until it is sufficiently decayed that they can tear it apart by the powerful teeth. Since crocodilian jaws do not move from side to side, they often must twist their whole bodies in an effort to detach a piece of flesh from a carcass.

A single species, the false gavial, makes up the family Gavialidae. Found in India, Bangladesh, Burma, and Pakistan, it is distinguished by its long, thin snout and rather bulbous head. The family Crocodylidae comprises all the other members of the order. These include the Nile crocodile, found widely in Africa; the mugger, which lives in India and Pakistan; and the American crocodile, which occurs in the southeastern United States and south to Ecuador. As in most other crocodilians, the bony plates occur only on the back and tail.

The American alligator and the Chinese alligator are the only two species of alligators. They were once much hunted for their skins and are quite scarce, although in some areas of the United States, they are now protected and making a comeback. The caymans of Central and South America are closely related to alligators but are quicker. They are mainly distinguished by the scutes, which cover both the upper and lower body.

The tuatara

The tuatara is the only surviving member of the order Rhynchocephalia, a group of reptiles that flourished more than 200 million years ago. This lizardlike animal is found only on a few islands off the coast of New Zealand. It preys on insects and other invertebrates and inhabits the burrows of sea birds such as shearwaters and petrels. It becomes active at night and seems able to function at lower temperatures than most reptiles.

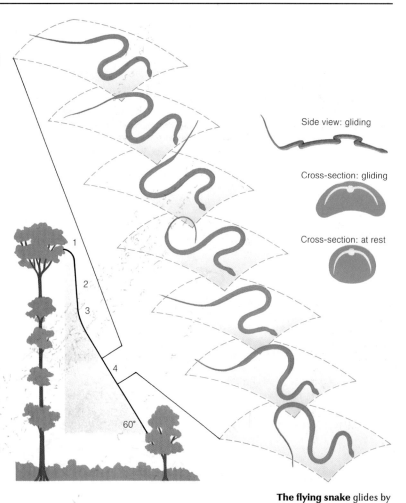

The flying snake glides by (1) projecting itself from a tree, (2) falling for several feet while (3) concaving its lower surface, and (4) undulating its body in a series of tight S-curves to gain some lift to slow the descent.

Crocodilian teeth are sharp and cone-shaped. In the crocodiles, a tooth on each side of the lower jaw fits into a pit in the upper jaw. This pit is open on one side, and the tooth protrudes when the mouth is closed, giving the characteristic "crocodile smile." One of their most surprising feats is that after the young crocodiles have hatched, the mother carries them gently to safety in her mouth.

Introduction to birds

Birds are relative newcomers to the animal kingdom. The earliest fossil recognizable as a bird, *Archaeopteryx,* lived about 150 million years ago during the Jurassic Period in a world dominated by reptiles. After *Archaeopteryx,* of which a few specimens and some feathers have been found, the next fossil birds—Hesperornis and Ichthyornis—come from the Cretaceous Period, between 138 million and 63 million years ago. *Hesperornis* was a water bird that lost the power of flight. *Ichthyornis* was a flying sea bird.

The first representatives of the modern families of birds appeared as early as the Eocene Period, between 55 million and 38 million years ago, and species alive today emerged during the Pliocene, between 5 million and 2 million years ago. The key to the success of modern birds lay in their development of feathers, flight, and warm-bloodedness. Birds now live in every habitat on earth, from polar regions to tropical deserts. Some have also adapted to life on water, thus conquering three environments—air, land, and water—something no other vertebrates have been able to do. Several groups of birds have lost the power of flight and have developed large size and strong legs to escape enemies.

The lightest skeleton

The bird skeleton has the same basic parts as that of other vertebrates, but it is extensively adapted for flight. The whole skeleton has become extremely light. The teeth have been replaced by a lightweight, horny bill. Many of the bones, like those of the skull, are very thin, and others, such as the limb bones, have a honeycomb structure—hollow, with thin internal struts that provide strength and rigidity without adding a lot of weight. The vertebrae in the backbone near the pelvic girdle are fused together to form the synsacrum, which provides firm support for the legs and cushions the bird against the shock of landing. To make up for this rigidity, the neck is extremely flexible, with up to 28 vertebrae—many more than in the neck of mammals, which have only 7 vertebrae.

Although the basic structure of a bird's limb bones is like that of most other vertebrates, there are some significant differences. For example, the femur (thighbone) of a bird is normally hidden because it is held up close to the body beneath the feathers. And what looks like the thigh of a bird is really a structure made up of fused shinbones, called the tibiotarsus. The equivalent to human shins are the elongated and fused bones of the ankle and feet, which form the tarsometatarsus. A bird foot typically has four clawed toes, upon which it walks. The arrangement of tendons and muscles in a perching bird ensures that as it bends its legs to perch, the toes are automatically pulled inward and the foot grasps the perch tightly. This means that a sleeping bird can remain attached to a perch without effort.

The collarbones (clavicles) are fused into a V-shape, forming the furcula, or wishbone. Together with the coracoids (shoulder blades) they form the pectoral girdle. This structure prevents the ribs from being crushed by the powerful wing muscles, which may make up as much as 30 per cent of a bird's weight. The pectoral girdle is attached firmly to the large breastbone, or sternum, which is shaped like the keel of a ship and which provides a strong anchorage for the wing muscles.

The forelimbs have been modified to form the wings, which are joined to the pectoral girdle. The humerus (bone of the upper arm) is short and stout and has a large surface area for the attachment of the flight muscles from the sternum. The ulna of the forearm is flattened to accommodate the secondary flight feathers. There are two wrist bones, three hand bones—two of which are fused together—and three finger bones. Attached to the first of the finger bones are the primary wing feathers. A flat membrane of skin is on each side of the wing bones. Together with long flight feathers, the broad wing surface helps create lift and power in flight.

Unique breathing system

A bird's lungs are connected by tubes to numerous thin-walled air sacs. These air sacs can take up one-tenth of the volume of the body, spreading into the spaces between the muscles, body organs, and even the hollow bones. The single-direction airflow system allows the bird to extract oxygen from the air even at such high flying altitudes as 25,000 feet (7,620 meters), where oxygen is in short supply.

Digestive system

Unlike almost all mammals, birds have no teeth. So instead of chewing food, birds break

A reconstruction of *Archaeopteryx,* a prehistoric bird that lived about 140 million years ago, reveals that, unlike modern birds, it had teeth and clawed fingers on the front of its wings. Feathers fringed its long tail, and it was probably capable only of short, clumsy flights from tree to tree. Since that time, birds have continued to evolve and adapt, so that today their representatives occupy nearly every terrestrial and semiaquatic habitat on earth. But above all, they have conquered the air.

it up with their bill or swallow it whole. Their digestive system features a thin-walled, baglike swelling called a crop at the base of the gullet, where food is stored and moistened before it passes into the two-part stomach. In the first part of the stomach, food is broken down by enzymes. In the second part of the stomach, which is called a gizzard, thick muscular walls grind up the food, sometimes with the aid of swallowed gravel or other coarse material. Food then passes into the intestines, where the nutritious matter and most of the water from the food is absorbed. Wastes are further processed and then passed into a chamber called the cloaca. Both the digestive system and the reproductive system connect to the cloaca. Wastes pass out of the bird through a single opening called the vent. Birds conserve water by eliminating waste in the form of solid uric acid.

The red-tailed hawk is found in North America and Central America. Belonging to the buzzard family, the red-tailed hawk is a superb flier and accomplished hunter. In many ways, it represents the pinnacle of bird evolution.

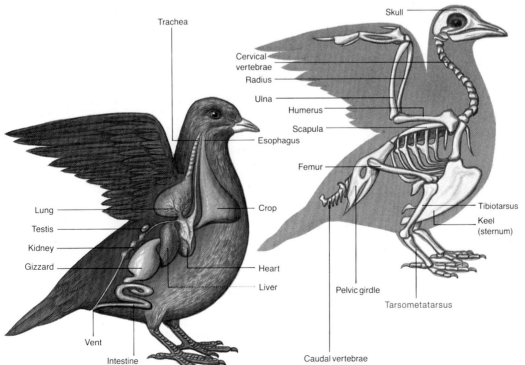

Trachea
Cervical vertebrae
Radius
Ulna
Humerus
Scapula
Esophagus
Femur
Skull
Crop
Tibiotarsus
Keel (sternum)
Lung
Testis
Kidney
Gizzard
Heart
Liver
Vent
Pelvic girdle
Intestine
Tarsometatarsus
Caudal vertebrae

The skeleton of a bird is designed to be lightweight and to provide adequate attachment points for powerful flight muscles. The main supporting structure is fairly rigid. But the whole skeleton is extremely light because nearly all the bones are hollow. Birds have a system of internal organs similar to most vertebrates, except that in many birds the digestive tract includes a crop, in which hard food is stored and moistened. Food passes from the crop into a gizzard, where it is ground. And, unlike other vertebrates, the urinary tracts and the lower intestinal tract come together in one chamber called the cloaca.

Classification of birds

There are more than 9,700 species of birds alive today, compared with only about 4,500 species of mammals. Apart from the nocturnal species, most are fairly easy to observe. Probably for this reason, bird watching is an increasingly popular pastime. As a guide to the enormous variety in the world of birds, this article outlines the makeup of each of the living orders—although not all taxonomists agree on the exact family groupings that constitute separate orders.

The orders of birds

Birds (class Aves) are classified into two subclasses, Archeornithes, containing the extinct *Archeopteryx*, and Neornithes containing all other birds. The Neornithes consists of four superorders: Odontognathae, or extinct toothed birds; Ichthyornithes, which are also extinct; and Neognathae and Impennes, which together contain all other birds. Birds are generally arranged in 28 orders and subdivided into 158 families. These orders are described in the summaries that follow. Scientists base this classification of birds on anatomy, behavior, and life history.

Order Struthioniformes. The ostrich, the largest living bird, is the single species in this order. It is a large, fast-running, flightless bird,

found today in the wild only in Africa. Several other orders of birds also have evolved a flightless life style. All have breastbones that have lost the keel because there is no longer any need for attachment of flight muscles. Together with the ostrich, these birds with flat breastbones are known as the ratites (from the Latin *ratis,* meaning raft). All other birds, which have keel-shaped breastbones are called carinates (from the Latin *carina,* meaning keel). The wings of the ratites are very small and the feathers are loose and fluffy, like the down feathers of young birds. They also have strong taut legs for fast running. The ratites include the emu of Australia and the cassowaries of Australia and New Guinea (order **Casuariiformes**); the kiwis of New Zealand (order **Apterygiformes**); and the rheas of South America (order **Rheiformes**). Two extinct orders, the moas of New Zealand (order **Dinornithiformes**) and the elephant birds of Madagascar (order **Aepyornithiformes**) were also ratites.

Order Tinamiformes. The tinamous are a group of about 50 species of birds that resemble game birds, but are probably related to the rheas. Tinamous have beautifully patterned feathers that help camouflage them from predators. They are found exclusively in South America and Central America.

Order Sphenisciformes. The penguins are found only in the Southern Hemisphere, in an area that extends from near the South Pole to the Galapagos Islands on the equator off the coast of Ecuador. Like the ratites, they have abandoned flight. Superbly adapted for fast underwater swimming, their wings are paddlelike, their bodies are stocky, and their feet are webbed. They eat fishes, squids, and crustaceans and generally breed in huge colonies.

Order Gaviiformes. The divers or loons are a primitive group of five species of water birds found only in the colder parts of the Northern Hemisphere. Their streamlined shape, paddlelike feet, and long necks and bills make them well equipped for catching fish under water.

Struthioniformes
Ostrich
Struthio camelus

Rheiformes
Common rhea
Rhea americana

Apterygiformes
Brown kiwi
Apteryx australis

Tinamiformes
Crested tinamou
Eudromia elegans

Casuariiformes
Emu
Dromaius novaehollandiae

Sphenisciformes
Magellan penguin
Spheniscus magellanicus

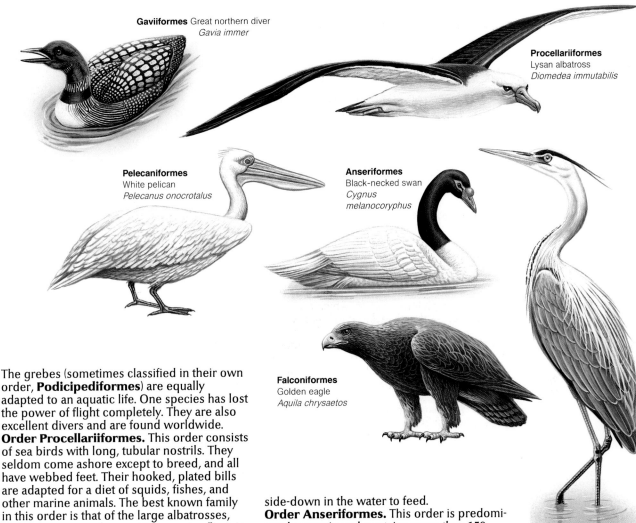

Gaviiformes Great northern diver
Gavia immer

Procellariiformes
Lysan albatross
Diomedea immutabilis

Pelecaniformes
White pelican
Pelecanus onocrotalus

Anseriformes
Black-necked swan
Cygnus melanocoryphus

Falconiformes
Golden eagle
Aquila chrysaetos

Ciconiiformes
Blue heron
Ardea herodias

The grebes (sometimes classified in their own order, **Podicipediformes**) are equally adapted to an aquatic life. One species has lost the power of flight completely. They are also excellent divers and are found worldwide.

Order Procellariiformes. This order consists of sea birds with long, tubular nostrils. They seldom come ashore except to breed, and all have webbed feet. Their hooked, plated bills are adapted for a diet of squids, fishes, and other marine animals. The best known family in this order is that of the large albatrosses, which are among the most impressive fliers in the bird kingdom. Other families are the shearwaters and petrels, the storm petrels, and the diving petrels.

Order Pelecaniformes. The pelicans are a group of large, fish-eating water birds and are the only birds with all four toes webbed. They are recognizable by their huge bills and highly flexible throat pouches for scooping fish out of the water. The other families are the gannets, or boobies, which dive vertically into the sea for fish from a height of 100 feet (30 meters) or more; the tropicbirds, graceful sea birds with long tail streamers; the 30 or so species of cormorants and shags, which are long-necked, long-billed, mainly black diving birds; the anhingas, or darters, which have long, snakelike necks and bodies; and the frigate birds, which are soaring birds that frequently pirate food from other sea birds.

Order Ciconiiformes. This order consists of the large, long-legged wading birds. The 50 or so species of herons, egrets, and bitterns, and the 16 species of storks in this order have daggerlike bills for spearing fish and other prey. The 30 species of ibises have slender, downcurved bills with sensitive tips for probing in mud, whereas the 6 species of spoonbills have spoon-shaped bills, which they hold open to trap small animals in the water. The 4 species of flamingos have hooked bills equipped with comblike plates that strain tiny organisms from the water pumped through them by the fleshy tongue. Flamingos hold their heads up-side-down in the water to feed.

Order Anseriformes. This order is predominantly aquatic and contains more than 150 species. The birds of this order range in size from the mute swan, which weighs 33 pounds (15 kilograms) and is one of the heaviest flying birds, to the tiny ringed teal, which weighs only 10.5 ounces (300 grams). The feet are webbed, and the bill typically is broad and flattened with fine plates at the edges for straining food from the water. They breed on every continent and major island in the world except Antarctica.

Order Falconiformes. This order comprises the birds of prey. It is a large order with almost 300 species, including the carrion-eating vultures and condors. All falconiformes have powerful, sharp, hooked beaks for tearing flesh, and strong feet with sharp talons for catching prey. They hunt by day, using their keen eyesight. Some, like the kites, eagles, and buzzards, use soaring or slow-flapping flight to spot their prey on the ground. Others, like the falcons, are fast fliers, catching birds and insects on the wing. Still others have special flying techniques for catching or feeding on fish, snakes, snails, and fruit.

Order Galliformes. Game birds belong to this order. The 10 species of megapodes, which are found exclusively in Australia and Indonesia, are medium-sized, brownish birds known for building huge mounds of soil and vegetation over their eggs. The fermentation of the organic matter in the mounds produces the heat that incubates the eggs. The 18 species of grouse in this order inhabit the temperate and

Galliformes
Spruce grouse
Canachites canadensis

Gruiformes
Crowned crane
Balearica pavonina

Columbiformes
White-winged dove
Zenaida asiatica

Charadriiformes
American oystercatcher
Haematopus palliatus

Psittaciformes
Rainbow lorikeet
Trichoglossus haematodus

Arctic regions of the Northern Hemisphere. Growing to about the size of a large chicken, they are plump with short bills and legs. There are some 35 species of pheasants, including the red jungle fowl of southern Asia and the East Indies, the peacock, and the smaller partridges and quails. The seven species of guinea fowl live in open country in Africa. The two species of turkeys are found in the woodlands of North America and Central America.

Order Gruiformes. This order includes the ground-nesting, ground-feeding cranes and rails. Many of this order are poor fliers, although the migratory cranes are a striking exception. Large, long-legged, and ranging in color from white to dark gray and brown, the 15 species of cranes have long secondary wing feathers drooping down over the tail. Some have ornamental tufts of feathers on their heads, which are used in their spectacular mating dances. Cranes inhabit marshy areas in Europe, Asia, North America, and Australia. The rails are among the most widespread of all land birds, and are found on all continents. They have also colonized remote oceanic islands. Most island rails have become flightless. They range in size from species as small as sparrows to birds as big as ducks. There are about 130 species of rails. They have long, narrow bodies; short wings and tails; and long legs and toes. The rail group includes the aquatic coots and gallinules.

Order Charadriiformes. This order consists of birds typically found on or near seacoasts and fresh water, including waders, gulls, and their relatives. The waders are a huge group of more than 200 species, ranging from the crow-sized oystercatchers and larger curlews to the medium-sized plovers and diminutive stints. Most are long-legged, and many have long bills for probing for food in mud. The skuas are dark-plumaged, gull-like sea birds with webbed feet, and most have elongated central tail feathers. They feed on fish, small mammals, birds and bird eggs. They also chase gulls, terns, and other birds and force them to disgorge food for their own consumption. There are five or six species. The gulls are a wide-ranging group that includes 43 species—from the large, great black-backed gull that grows to about 30 inches (75 centimeters) in length to the 10-inch (25-centimeter) little gull. Gulls are typically grayish or brownish when immature and gray-and-white or black-and-white as adults. Most gulls are coastal, but some live inland. Many have been able to expand their range dramatically, because they feed on food thrown away by humans. Terns are slender sea birds with short legs, webbed feet, and long, pointed bills. They occur in white, gray-and-white, black, or black-and-white and are found worldwide. The auks are short-winged black-and-white diving birds of the northern oceans. Although they can fly, they are most at home in the water. Auks are the northern equivalent to the penguins, also nesting in huge colonies. They range in size from the 6-inch (15-centimeter) little auk to the 30-inch (75-centimeter) extinct great auk. The 22 living species include the familiar puffins.

Order Columbiformes. This order contains the pigeons and their relatives, and includes the extinct dodo. There are 300 species of pigeons and doves found throughout the world, except for Antarctica. They live on seeds, fruits, and berries. Whereas most birds drink by drawing water into the mouth and lifting the head back to swallow, these birds can suck in water without lifting their heads.

Order Psittaciformes. This order consists of approximately 315 species of parrots, which live primarily in the tropics. Varying in size from 3 inches (8 centimeters) to more than 3 feet (91 centimeters) long, most parrots are mainly green, although some are brilliantly colored, and many have long tails and stout, short, strongly hooked bills adapted for feeding on fruits, berries, and seeds.

Order Cuculiformes. This order includes the 19 species of brightly colored touracos from the jungles of Africa and the cuckoos. Some of the cuckoos are parasitic breeders, laying their eggs in the nests of other birds and allowing them to rear their young. More than 100 species of cuckoos are found worldwide.

Order Strigiformes. This order is made up of the owls. Most owls are nocturnal birds of prey, with large eyes at the front of their heads; hooked, flesh-tearing bills; razor-sharp talons; and dense, soft plumage, which helps make them almost noiseless at night. They are subdivided into the 10 species of barn owls and bay owls and the 121 species of typical owls. Owls are found throughout the world, except for Antarctica.

Order Caprimulgiformes. This order con-

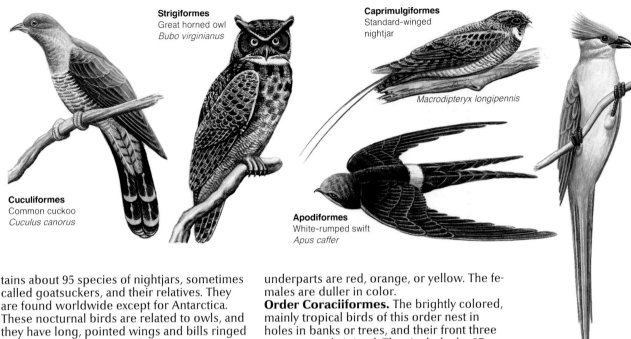

Strigiformes
Great horned owl
Bubo virginianus

Caprimulgiformes
Standard-winged
nightjar

Macrodipteryx longipennis

Cuculiformes
Common cuckoo
Cuculus canorus

Apodiformes
White-rumped swift
Apus caffer

Coliiformes
Blue-naped mousebird
Colius macrourus

tains about 95 species of nightjars, sometimes called goatsuckers, and their relatives. They are found worldwide except for Antarctica. These nocturnal birds are related to owls, and they have long, pointed wings and bills ringed with bristles for trapping flying insects.

Order Apodiformes. All the members of this order, which includes the swifts and hummingbirds, are exceptional fliers with pointed, slender, but powerful wings and tiny feet that are useless for walking. The swifts are the most aerial of all birds, able to sustain flight for many hours. There are about 75 species of swifts and they are found virtually throughout the world. All of the more than 300 known species of hummingbirds live in the Western Hemisphere. Beating their wings as fast as 70 times per second, hummingbirds can hover and even fly backwards. Most are tiny and the largest is only the size of a sparrow. They feed on nectar and insects.

Order Coliiformes. This order consists of the single family of coly, or mousebirds. All six species are African and are small birds with very long, stiff tails and prominent crests.

Order Trogoniformes. This order contains the brightly colored, tree-dwelling, tropical trogons, which have unusually delicate skins. Dark, metallic-blue, -green, or -violet feathers cover the head and back of most males. Their underparts are red, orange, or yellow. The females are duller in color.

Order Coraciiformes. The brightly colored, mainly tropical birds of this order nest in holes in banks or trees, and their front three toes are partly joined. They include the 87 species of kingfishers, 8 species of motmots, approximately 25 species of bee-eaters, 16 species of rollers, 8 species of wood hoopoes and hoopoes, and 44 species of huge-billed hornbills.

Order Piciformes. Birds in this order have feet with two toes pointing forward and two backward. They include the approximately 200 species of woodpeckers, which are specialized for feeding and nesting in tree trunks, and about 40 species of fruit-eating toucans of the tropical rain forests of Central America and South America.

Order Passeriformes. This order, the perching birds, is the biggest group of all, containing more than a third of all living families and more than half the living species. All have feet adapted to perching on or clinging to branches or other supports. The order includes the "song birds," which have developed the ability to sing. It also contains the swallows, wagtails, pipits, wrens, thrushes, warblers, tits, finches, weavers, sparrows, starlings, and crows.

Trogoniformes
White-tailed trogon
Trogon viridis

Coraciiformes
Common kingfisher
Alcedo atthis

Piciformes
Ivory-billed
woodpecker
*Campephilus
principalis*

Passeriformes
Wood thrush
*Hylocichla
mustelina*

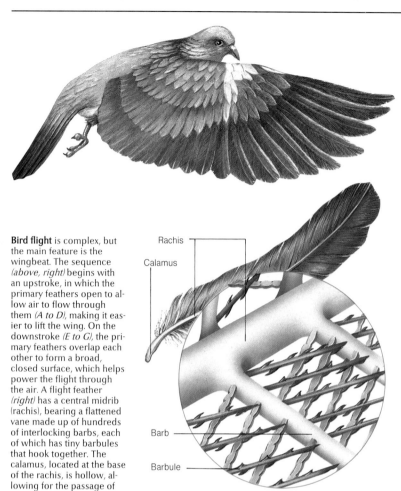

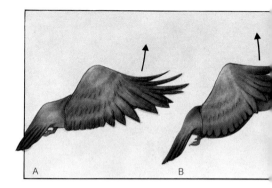

Bird flight is complex, but the main feature is the wingbeat. The sequence *(above, right)* begins with an upstroke, in which the primary feathers open to allow air to flow through them *(A to D)*, making it easier to lift the wing. On the downstroke *(E to G)*, the primary feathers overlap each other to form a broad, closed surface, which helps power the flight through the air. A flight feather *(right)* has a central midrib (rachis), bearing a flattened vane made up of hundreds of interlocking barbs, each of which has tiny barbules that hook together. The calamus, located at the base of the rachis, is hollow, allowing for the passage of nutrients to the feather.

Rachis

Calamus

Barb

Barbule

Feathers and flight

The two factors that set birds as a class apart from nearly all other vertebrates are feathers and flight. Only bats, which are mammals, are also capable of powered flight. Feathers provide insulation and the lifting surface of the wings. The mobility provided by flight has enabled birds to spread and establish themselves in nearly every habitat on earth.

Feathers

Bird feathers are made of the protein keratin, the same basic material as their horny bills and the scales on their legs. Each feather develops from a knob (papilla) within a feather socket, or follicle. The follicles are arranged in distinct areas called feather tracts over the bird's body. Each follicle produces one, two, or even three feathers every year.

There are two main types of feathers—pennae and plumulae. The pennae are the flight and contour feathers. A typical penna has a broad, flat vane attached to both sides of a long central shaft. The shaft consists of two parts. A hollow, rounded base, called the calamus or quill, extends from the bird's skin to the vane. The solid, tapering upper part of the

shaft, called the rachis, runs through the vane. The vane is formed by barbs that branch out from the sides of the rachis and barbules that branch from the barbs. Hooks on the barbules link neighboring barbs, giving the vane both strength and flexibility.

The flight feathers (primaries and secondaries) are directly concerned with flight, whereas the contour feathers have several functions. They give the bird its streamlined shape, helping it to cut through the air efficiently. They also provide vital insulation by trapping a layer of heat-retaining air close to the skin. During hot weather, a bird opens up its contour feathers to allow heat to escape, and in cold weather it fluffs them up to increase the insulating layer of air.

Another function of feathers is to waterproof the bird. They give the bird its color and may be used for camouflage, for species recognition, for sexual display, or as warning signals.

The plumulae are the down feathers. They lie beneath the contour feathers, providing extra insulation, and are the only feathers on a newly hatched chick. Down feathers are much simpler than pennae, with a very short midrib and no barbules, so that the barbs are separate, giving them their fluffiness.

The number of feathers varies considerably from one species to another. Usually, the larger the bird, the more feathers it has. For example, a tiny hummingbird may have fewer than 1,000 feathers, whereas a large swan may have more than 25,000. The number of feathers also varies from one season to another.

Because feathers are subject to great stress and can wear out, all birds molt at least once a year. Most birds molt their feathers gradually, so there is no interference with flight. But ducks lose their flight feathers all at once and so experience a period when they are flightless and particularly vulnerable.

Despite their proverbial lightness, all the feathers on a bird are, surprisingly, often heav-

The shapes of birds' bills reflect adaptations for dealing with different types of food. The strong, hooked bills of cormorants and eagles tear the flesh of their prey, whereas those of macaws and toucans crack nuts and open fruit. A gull's "all-purpose" bill suits its role as a scavenger, and a duck uses its flattened bill to strain food from the water. A blue tit's short bill catches insects and grubs, and the extremely long, narrow bill of the hummingbird allows it to sip the nectar deep inside tubular flowers.

ier than the bird's incredibly light, hollow-boned skeleton. The feathers of the bald eagle, for instance, weigh more than 1.5 pounds (670 grams), but the skeleton weighs only about 9.5 ounces (270 grams).

Flight

Bird flight is governed by the same laws of aerodynamics that control all heavier-than-air flight. There are two opposing forces involved in the mechanics of bird flight: lift, the upward force that keeps the bird airborne, and drag, the force of the air opposing lift, which is largely a result of air turbulence at the wingtips.

A bird's wing is rounded above and hollow below, with the leading edge thicker than the trailing edge—a shape that produces maximum lift. The convex upper surface causes the air flowing over it to travel faster than the air flowing past the undersurface. As a result, the pressure on the underside of the wing is greater than that on the upper surface, producing lift.

The angle that the wing makes with the horizontal is called the angle of attack. The greater the angle of attack, the more the airflow over the wing becomes turbulent and the more the bird is likely to stall. At the optimum angle for flight, upward lift counteracts the forces of gravity and drag and keeps the bird aloft.

The main function of the primary feathers is to provide forward thrust. The highly flexible primary feathers of the wingtips twist, acting almost like a propeller that drives the bird through the air. The wingtips move faster than the rest of the wing. The main function of the secondary feathers is to provide lift—they move very little as the wing beats. The frequency of wingbeats varies greatly, from about two beats per second in a swan to 50 or more per second in a hummingbird.

To alter the direction of flight, a bird increases the angle of attack of one wing, or

beats that wing faster, creating a difference in lift on the two sides of its body. Movements of the tail are also used. To land as gently as possible and minimize the shock to its body, a bird must be on the point of stalling. To achieve this, it pushes its body downward and spreads its wings and tail.

Flight speeds of birds also vary, although typical speeds are between 20 and 35 miles (32 and 56 kilometers) per hour. The fastest bird is probably the peregrine falcon, which dives on its prey at speeds of 180 miles (290 kilometers) per hour or more.

Some birds save energy by gliding instead of flapping their wings. Vultures, for instance, use their long, broad wings to soar in thermals (rising currents of warm air). Albatrosses tack back and forth low over the waves with their long narrow wings, using the updrafts produced by the friction of the air with the water to stay aloft.

Hovering in midair, a tiny ruby-throated hummingbird uses its long bill to feed on nectar from sage flowers. The wings beat at more than 70 times a second, requiring its heart to beat 600 times a minute and using five times as much oxygen per ounce of body weight than a larger, slow-flying bird.

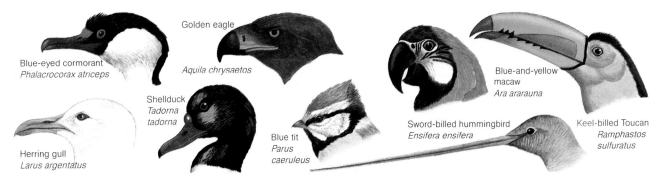

Blue-eyed cormorant
Phalacrocorax atriceps

Golden eagle

Aquila chrysaetos

Shellduck
Tadorna tadorna

Herring gull
Larus argentatus

Blue tit
Parus caeruleus

Sword-billed hummingbird
Ensifera ensifera

Blue-and-yellow macaw
Ara ararauna

Keel-billed Toucan
Ramphastos sulfuratus

In a courtship display, a male great frigate bird spreads his wings and inflates his conspicuous scarlet throat-sac. After mating, the female lays a single white egg, which she incubates for six to seven weeks. The chick is dependent on its parents for food from 6 to 10 months, spending the first 4 or 5 months in the nest.

Breeding and behavior in birds

Birds live in all parts of the world, in nearly every habitat—in forests, deserts, cities, grasslands, on mountaintops and islands, and even in caves. Although this range of habitats has led birds to develop a great variety of behaviors, all birds share certain characteristics. All birds hatch from eggs, which, in most species, are laid in a nest that is built by one or both of the parents. Most baby birds remain in the nest for several weeks or months after hatching, and the parents feed and protect them until they can fend for themselves.

Territory

Most birds breed within a definite territory, an area that they defend against rivals. The territory of many perching birds is about 1 acre (0.4 hectare) in extent, but in large birds of prey—which get food far less frequently over a much wider area—it may be as much as 29 square miles (75 square kilometers) in area. Territory helps birds to survive by parceling out the available habitat into areas capable of supporting a pair of birds, allowing them to feed and breed there without competition from other birds of their species.

Courtship

Birds recognize potential mates of the same species and breed with them following courtship. A male and female form a relationship called a pair bond after the male performs a series of courtship displays and the female responds favorably, often in the form of another display. The type of courtship behavior varies, as does the type of pair bond formed. Many species are monogamous and establish lasting bonds, but others are polygamous, with successful dominant males mating with several females.

A vital part of bird displays is the emphasis on the colors and patterns that distinguish a particular bird from otherwise similar species. Where several closely related species coexist, as with the ducks of the Northern Hemisphere, the plumage of the males tends to be distinctive, differing strikingly between species. On the other hand, in areas with only one species, the plumage can be indistinctive. In places where similar-looking birds of different species coexist, as in the European and Asian woodlands of the virtually indistinguishable willow warbler and chiffchaff, distinctive songs identify the birds.

Another important function of courtship is to encourage the male and female to reduce territorial hostility and accept each other as mates—a process that may take several weeks. For example, the male European robin sings at the boundaries of his territory, both to warn off rival males and to attract potential mates. Sooner or later, his song encourages a female to approach him. The male first responds by adopting a threat display, using his red breast as a "flag" to discourage the intruder, just as he would if a rival male was involved. A rival male would either return the threat or flee. But the female does neither, adopting a submissive posture instead. This gradually encourages the male to allow her into his territory, and eventually to mate, nest, and rear a family.

Unlike the robin, some male birds do not display in an individual territory, but come together at communal display grounds, or leks. Game birds, such as the American grouse or the capercaillie of northern European and Asian coniferous forests; certain waders, such as ruffs; and birds of paradise, all have leks. The females go there only to pair and mate. They take no part in the display, leaving this entirely to the colorful males.

Not all birds have displays in which the male performs or sings and the female takes on a submissive role. Many species have courtship rituals that involve both partners equally, including the strange, complex dances of grebes, the elaborate bill-scissoring displays of some albatrosses, and the wild dances of many cranes.

Displacement activities

Courtship ceremonies among birds are usually modified and highly stylized versions of everyday actions, such as preparing for flight, preening, drinking, or feeding. Courtship is a tense period, involving the the sex drive and the conflicting emotions of aggression and fear. A bird may therefore indulge in ritualized behavior that alternates between a desire to flee and a desire to approach the mate—the "fight or flight" situation. If the conflict is great enough, as when a female behaves aggressively toward a male, the bird may show "irrelevant" behavior, such as beak-wiping. This is called displacement activity. In many birds, displacement activities have become ritualized and form an important part of their courtship display. Other courtship ceremonies originate from the behavior of young birds. For example, female finches flutter their wings and beg food from the male before mating.

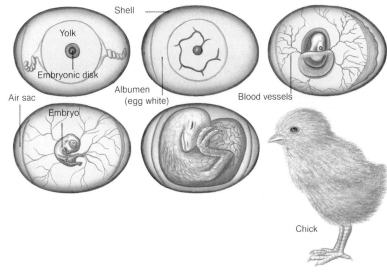

Development of a bird embryo takes place within the egg during incubation. Most nutrients are contained in the yolk, with calcium for bone growth coming from the shell. The shell is porous and allows the passage of gases, such as oxygen and carbon dioxide.

Mating

Successful copulation, resulting in fertilization of the egg or eggs, depends on the full cooperation of the female. Unlike mammals, most male birds do not have a penis. Sperm is transferred from the cloaca of the male to that of his mate. To encourage better contact between them, the female sticks out her cloaca and moves her tail to one side.

Mating takes place mainly during the time when the female is carrying unfertilized eggs. The single egg of the Emperor penguin needs only one fertilization, whereas birds such as songbirds, which lay several eggs, may involve several fertilizations.

Nests and nesting

Birds vary enormously in their choice of nest sites, nesting materials, and methods of nest construction. Some penguins, for instance, build no nest at all. The emperor penguin incubates its single egg on its feet, where it is warmed by a feathered flap of belly skin for nine weeks during the subzero temperatures and perpetual darkness of the Antarctic winter. Most waders nest in a hollow on bare ground, generally lining the hollow with gravel or a few twigs. Some species of gannets, auks, and gulls nest on virtually inaccessible sea cliffs where they are safe from mammal predators. The eggs of auks, which nest on particularly perilous cliff edges, are pearshaped. As a result, they roll around in circles instead of over the edge of the cliff.

Many birds, such as owls, woodpeckers, and hornbills, nest in holes in trees. Kingfishers, some swallows, and bee-eaters nest in holes in sandy banks. Such birds generally lay white eggs, since there is no need for camouflage in these situations. Others, such as herons, crows, and many birds of prey, nest high up in tall trees, building strong, bulky nests of branches and large twigs. Some of the cave swiftlets of

Contrasts in nest-building are demonstrated by herons and weaverbirds. The great blue heron *(left)* lives in Canada and the United States and builds ragged nests of twigs in colonies perched on tall trees and buildings. The black swamp weaver *(right)* builds a delicate nest in reeds by interlacing lengths of grass to form a globular home with an entrance near the top.

Camouflaged coloration is a significant feature of ground-nesting birds and their eggs. During the breeding season, the speckled back plumage of the golden plover makes it almost invisible.

A small dunnock, or hedge sparrow, tries to keep up with the voracious appetite of its enormous "offspring," a cuckoo chick, which has pushed the dunnock's own eggs out of the nest. The adult cuckoo had laid its egg in the nest to be hatched and cared for by the dunnock.

ments, red-brown and blue-green, give eggs their great range of colors and patterns. White eggs contain no pigments.

Eggs vary in shape from the almost spherical eggs of owls to the long, pointed eggs of swifts, waders, and guillemots. They also vary enormously in size, from the tiny eggs of hummingbirds, which can be as small as only 0.2 ounce (5.7 grams), to the huge ones of ostriches, which are 8 inches (20 centimeters) long and weigh 3 pounds (1.4 kilograms).

Unlike those of fishes and amphibians, the eggs of birds are adapted to allow the embryo to develop on dry land. A bird's egg is a closed system. The yolk feeds the embryo, and the tough but porous shell provides protection yet also allows excess water to pass out of the shell and air to pass into the shell for the embryo to breathe. The chick excretes uric acid, which does not dissolve in the body fluids and cause harm. This waste is secreted into the baglike allantois, which is left in the shell after the chick hatches.

Incubation

Unless birds eggs are kept at a relatively high temperature—about 102° F. (39° C)—the chicks within them will not survive. To maintain the heat, most birds develop naked brood-patches on their abdomen, well supplied with blood vessels, which transfer their body heat to the eggs. Among most species, only the female sits on the eggs. In some birds, both sexes share the task of incubation. In a few, such as cassowaries and the emu, only the male incubates. The Australasian megapodes, or brush turkeys, do not brood. Instead, they build huge mounds of decomposing vegetation that generate enough heat to incubate the eggs.

Hatching

The chick breaks out of its protective prison by chipping out a series of holes in the middle of the shell with a sharp, hard bump called an egg tooth at the end of its upper bill. It then cracks the egg open by pushing in opposite directions with its head and feet.

Chicks are of two basic types. Precocial chicks can walk from the nest within a few hours or days of hatching. Some precocial chicks, such as the chicks of plovers, find all their own food from the start. They are well equipped with strong legs and a covering of warm down feathers. Their eyes are open and they are usually superbly camouflaged against predators. Examples of precocial chicks are the young of wild fowl, game birds, and waders. The second type of chicks are known as altricial. They hatch at a much more immature stage of development—when they are still blind, helpless, and practically featherless. Altricial chicks are totally dependent on their parents and remain in the nest until almost ready to fly. Examples of such chicks are the young of pigeons, owls, thrushes, hummingbirds, woodpeckers, and all songbirds.

Migration

One of the main reasons birds are so numerous and widespread in the world is that they are capable of moving far and fast. Many

Asia use their thick, gummy saliva to cement together their building material and glue the nests to the walls or roofs of caves. Such nests are collected for making bird's-nest soup.

The role of the sexes in nest-building varies greatly. In most birds, both the male and female are involved, but in finches and hummingbirds, among others, the female does most of the work. In some, such as weaverbirds, the males are the nest-builders.

Eggs and egg laying

After the male's sperm has fertilized the egg in the female's ovary, the egg—now consisting of an ovum (egg cell) plus the yolk—moves down the oviduct. It is covered by the jellylike "egg white," or albumen, then by two shell membranes, and finally by the shell itself, which consists of several layers covered by a thin cuticle. The pigments that give the egg its color are laid down mainly in the cuticle and outer shell layers. Mixtures of just two basic pig-

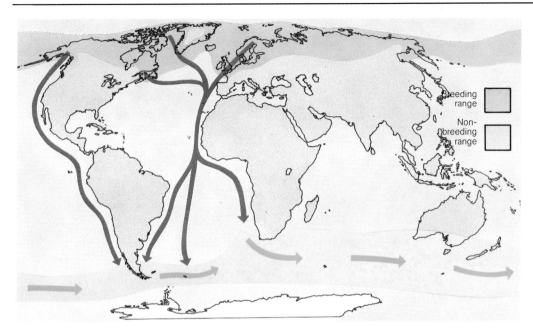

The farthest-traveling migrant bird, the arctic tern, flies annually about 22,000 miles (35,400 kilometers) from its breeding grounds in the north to the Antarctic and back again. For much of the year, it lives over the oceans, flying along the eastern shoreline of the Pacific Ocean or across the Atlantic. At the most northern and southern extremes of its range, the bird lives in perpetual daylight.

species make long migrations in search of food or breeding sites. True migration involves regular seasonal journeys between a breeding area and a wintering area, where the climate is more favorable and the birds can find food. Many insect-eating birds, such as swallows and warblers, breed in temperate regions and migrate to lower latitudes in autumn, to spend winter in the tropics. Others, including some wild fowl, waders, and some thrushes, breed in high latitudes and migrate to temperate regions for the winter.

Navigation

Many birds follow well-defined migratory flyways, which often parallel coastlines, mountain ranges and valleys, or ocean currents. Birds that migrate by day follow various landmarks. They do not need to be taught their route. Young Old World common cuckoos, for example, find their way back thousands of miles to their winter quarters in Africa completely unaided, their parents having returned several weeks before them. But the instinctive following of landmarks cannot explain the phenomenal journeys of sea birds that fly across the vast Pacific Ocean, where there are no clues to guide them. Nor can they explain how a Manx shearwater taken from the coast of Wales and flown in an aircraft to Boston in the United States found its way back to its nest over more than 3,000 miles (4,828 kilometers) of open ocean in 12 1/2 days—arriving before the letter announcing its release.

Much remains to be learned about how birds perform such astonishing feats of navigation. According to some experts, it is possible that they possess a sort of internal compass that enables them to use the sun by day and the stars by night to get their bearings. It also seems likely that certain birds are guided by the earth's magnetic field, especially when the sky is overcast.

Migrating geese fill the sky as they travel southward from their breeding grounds in arctic Canada. Despite the apparent confusion of the flock as they take off together, the geese soon group themselves into a characteristic V-shaped formation (see insert). In autumn, the white snow goose and its smaller blue subspecies, both of which appear in the larger photograph, travel 1,700 miles (2,700 kilometers) at an average of nearly 30 miles (50 kilometers) per hour to winter in the marshes fringing the Gulf of Mexico.

pets

Over the centuries, humans have developed special relationships with various other species of animals. One such association is the mutually beneficial relationship we enjoy with our pets.

Pets provide people with warm companionship. This young boy is enjoying the company of two guinea pigs and a Bernese mountain dog.

The domestication of animals

Besides fruits, vegetables, and grains, an important part of the human diet is the meat of other animals. Other foods we get from animals include eggs, milk, and honey. In addition, humans use animals for clothing materials like leather, silk, and wool. To make these products easier to get, people began to domesticate, or tame, certain animals at least 10,000 years ago.

Today, some domesticated animals provide food. Others provide labor—the water buffalo that plows an Asian rice field, for example, or the camel that carries people and goods across an African desert. Soon after they were domesticated, cats and dogs became important because they provided services for humans. Cats were tamed and kept in households to kill mice and rats. Dogs were used to help in the hunt and to protect people by warning them of approaching danger. Eventually, both species became not only domesticated animals but pets—animals kept more for emotional than economic reasons.

Humans and their pets

Dogs and cats have been kept as pets in all parts of the world for thousands of years. Fish and birds, such as parakeets and canaries, are also popular pets. Certain animals are considered pets in some parts of the world and not in others. People in Malaysia, for example, enjoy mongooses as pets. People in China make pets of cormorants.

Problems can arise when a person tries to keep a wild animal as a pet. A lion cub may be as emotionally appealing as a kitten, but the cub will grow up to be a full-sized lion, and it will never be anything but a wild animal. It may resort to instinctive aggressive behavior, such as clawing and biting, whenever such behavior would be appropriate in the wild. Its owners and other humans, especially young children, are then endangered. For this reason alone, many cities and towns have laws against keeping wild and exotic animals as pets in homes.

When the owners of such wild animals realize the mistake they have made—or simply change their mind about owning the animal—they find that zoos and animal shelters will generally refuse to accept their pet. The unfortunate animal may end up on a "canned hunt" ranch, where persons pay thousands of dollars to "hunt" them.

Other wild animals pose different problems as pets. Many species of birds, for example, outlive their owners. Parrots may live for 70 years. And every wild animal must be taken forcibly from the wild. Of the estimated 2 to 5 million birds taken from the wild each year, as many as 60 per cent may die while being transported.

Some owners of exotic pets are part of a U.S. government-sponsored program to protect endangered species by breeding the animals in captivity. These animals were not born in the wild, however. They were bred by individuals or organizations who are registered with the Fish and Wildlife Service's Captive-Bred Wildlife Registration program, established in 1979. The breeders hope to preserve these endangered species in case they become extinct in the wild. In some cases, the species may eventually be "reintroduced"—released—into the wild.

Small animals often make good pets for children. By caring for their pets, children learn responsibility.

Exotic animals may not be well suited for pet life. The macaw of Central and South America, for example, can live many years and some outlive their owners.

Introduction to Mammals

The class Mammalia is a relatively small one, containing about 4,500 species and 19 orders. Among the largest orders are the rodents, or Rodentia, which make up about half of all mammal species, and the bats, or Chiroptera, which account for about one-quarter of mammal species.

Mammals are one of the most diverse groups of animals, varying remarkably in structure, size, and habitat. The smallest living mammals are the shrews of the family Soricidae. The pygmy shrew, for example, weighs less than 1 ounce (28.35 grams). The largest mammal is the blue whale, with a weight of more than 220 short tons (200 metric tons).

Distinguishing features

Mammals are set apart from other animals by several characteristics. One of their most distinctive characteristics is their hair or fur, which in most mammals covers the entire body. Mammals are also endothermic and homeothermic—that is, they produce heat internally and maintain a constant body temperature. Their coats of hair help regulate their body temperatures by acting as an insulator to prevent loss of body heat. Mammals maintain a fairly constant body temperature that normally lies between about 92° F. (34° C) and 102° F. (39° C). This heat is produced in a process called metabolism, which produces the energy and substances that an organism needs to live.

Mammalian metabolism produces heat. This heat production is controlled by a region of the brain called the hypothalamus. In most mammals, the hypothalamus works along with sweat glands in the skin that are distributed over much of the body. When a mammal's body becomes too hot, the hypothalamus becomes active, causing blood vessels to dilate and release excess heat. The skin is cooled when sweat secreted onto the skin evaporates. When a mammal becomes too cold, the blood vessels contract, and the sweat glands dry up. In addition, reflex shivering occurs, which generates heat by muscle contractions. Such heat regulation allows mammals to remain active regardless of the temperature of their environment. In contrast, reptiles and other animals whose body temperature depends on the temperature of their environment cannot function in cold weather.

Other features of mammalian anatomy also set them apart from other classes of animals. The bones and teeth of mammals are designed for greater efficiency. For example, mammals have three small bones called ossicles in the inner ear. These bones help give them sensitive hearing. Their lower jaws are formed from a single bone, and a variety of teeth with different forms and functions and complex cusp patterns gives them the ability to chew their food more thoroughly. In addition, female mammals have mammary glands, or mammae, which produce milk to feed their young. Most mammals also have fleshy lips. When a mammal is young, it uses its lips to suckle, but as it grows, the mouth may become adapted to feed on specialized food.

Like fishes, birds, and crocodiles, mammals have a four-chambered heart and a diaphragm that separates the chest and abdominal cavities so that the lungs can work more efficiently in supplying increased amounts of oxygen to the blood.

Habitats

Mammals have been able to exploit an amazing variety of habitats. Moles (family Talpidae) and many rodents make their homes in underground burrows or among leaf litter on the ground, as do shrews. Otters (family Mustelidae) and beavers (family Castoridae) lead a semiaquatic life in fresh water. Whales and dolphins (order Cetacea) spend all of their life in the sea. Monkeys and sloths (orders Primates and Edentata) inhabit tree canopies in forests. Bats (order Chiroptera) live in the air at least part of the time. Polar bears (order Carnivora) and musk oxen (order Artiodactyla) live

Suckling piglets are typical of young mammals, all of which are initially nourished by mother's milk. The domestic hog often has large litters of between 6 and 14 piglets. Although the piglets are covered by fine velvety hair, adult domestic hogs, unlike most mammals, have hardly any hair, just a sparse scattering of bristles.

Hair, a distinguishing feature of most mammals, covers the body of this orangutan of Borneo. The main purpose of hair on mammals is to help prevent the loss of body heat.

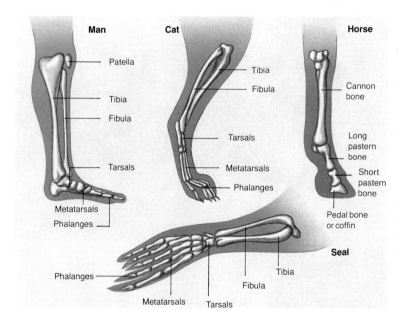

The limbs of mammals conform to a basic pentadactylic, or five-fingered, pattern. The lower limb consists of the heavy, weight-bearing tibia and the slighter fibula, and a five-toed foot containing the ankle bones (tarsae); metatarsals; and phalanges, which are the small bones that make up the toes. There are three phalanges on each toe except for the large innermost one. The structure of the foot varies among species. Plantigrade feet are found in humans and bears and are constructed so that the whole foot is placed on the ground when walking. Animals such as dogs and cats, which run fast, have digitigrade feet, in which only the part containing the phalanges is used for walking. Ungulates walk with unguligrade locomotion, on the tips of the last phalanges. In horses, the tibia and fibula have fused to form the cannon bone, and the tarsals have fused to form the pastern. The only remaining toe is the third one, which is the hoof. In aquatic mammals, such as seals, the foot has assumed a finlike shape.

in arctic environments, while camels (order Artiodactyla) and kangaroo rats (order Rodentia) live in desert conditions.

Teeth

Because mammals are warm-blooded, they consume more energy than cold-blooded animals. They therefore need to eat large amounts of food to fuel their metabolism. Mammals' complex teeth are specialized to cut, tear, and grind their food so that digestive processes can start earlier and are more thorough than if the food were processed only in the stomach.

Mammals' teeth closely reflect their diet. For example, cutting teeth in the front of the jaws, called incisors, are extremely well developed in the rodents, which need them for cutting through tough, woody materials. Carnivores have small incisors, but highly developed canines, which they use to tear and pierce flesh. They also have carnassials, which are modified forms of the last upper premolar and the first

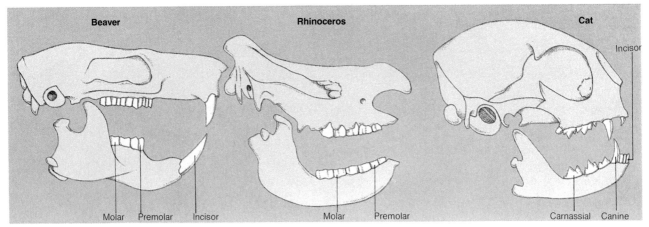

Beaver   Rhinoceros   Cat

Incisor

Molar Premolar Incisor   Molar Premolar   Carnassial Canine

The teeth of mammals often reflect their diet. The jaws of herbivorous rodents, such as the mountain beaver, lack canines and have very large upper and lower incisors to gnaw at woody vegetation. Their premolars and molars are almost flat and are therefore good for grinding vegetation. Grazing vegetarians, such as rhinoceroses, usually have no incisors or canines. Carnivores, such as cats, have small incisors, but large canines for tearing flesh. Their premolars and molars (carnassials) have sharp edges that can slice through meat.

lower molar. Carnassials are bladelike to slice through flesh and cut meat into bite-sized pieces by their scissor action. In herbivores, which graze on grasses or nibble on leaves, the flat-topped premolars and molars, used for grinding non-woody vegetation, are the most important teeth.

Some mammals, such as pangolins and anteaters, which feed on termites and ants, and baleen whales, which feed on shrimplike crustaceans, have no teeth at all. These mammals use special mechanisms for collecting and controlling their prey. The giant anteater, for example, uses its long, sticky tongue to sweep up insects from the ground. The anteater's digestive system includes a gizzard that grinds up the hard exoskeleton of ants and other invertebrates.

Movement

Most mammals are quadrupeds—that is, they move about on four feet. Some have plantigrade locomotion and walk by placing the whole foot on the ground. Plantigrade motion is usually found in species that hold or manipulate their food, such as primates and bears. Mammals with digitigrade locomotion walk on their toes, keeping their heels and palms off the ground. Digitigrade locomotion enables such animals as cats and dogs to run at high speeds. Such animals also have a limited ability to manipulate food or other objects. Animals with unguligrade locomotion, called ungulates, or hoofed animals, walk only on the tips of the toes, which are protected by large nails or hoofs. Unguligrade locomotion is associated with fast, long-distance travel, and the animals that use it typically run to avoid their predators. Ungulates have almost no ability to use their hands or feet to manipulate their food or environment.

Humans are the only animals to walk on two feet, or bipedally. But many rodents and marsupials, such as kangaroos, hop bipedally.

The limbs of aquatic mammals have evolved into paddlelike flippers or fins. Some, such as the otter and the platypus, have a membrane between the digits on the forelimbs, which aids them in swimming. In seals, all the flippers are used for moving through water. In whales, the hind limbs have completely disappeared and locomotion is powered by movements of the back and tail, whereas the func-

tion of the foreflippers is only to steer and balance.

Gibbons, orangutans, and other apes have strong arms and elongated hands that allow them to swing from branch to branch in a type of locomotion known as brachiation. Some South American monkeys also have prehensile tails that operate as a fifth limb. Some squirrels and marsupials that live in the tree canopy have folds of skin that extend from the forelimb to the hind limb and tail, on which they glide from tree to tree. The only mammals that are true fliers are the bats. Their forearm is the major support of a wing membrane. The hind limb and the tail often also support the wing.

Classification of mammals

The class Mammalia is divided into two subclasses: the Prototheria and the Theria. The prototherians are a group of primitive mammals. Today, only one order of prototherians exist—the Monotremata. The order has two families: the Ornithorhynchidae, or bird-noses, and the Tachyglossidae, or echidnas.

Monotremes have characteristics of both reptiles and mammals. Like reptiles, they are oviparous, or egg-laying. They are also toothless, though young platypuses produce three tiny teeth soon after birth, which they lose and replace with horny plates. Also, like reptiles, monotremes have only one opening at the hind end of the body. This opening, called the cloaca, serves both the processes of elimination and reproduction. Female monotremes have mammary glands but no teats, or nipples, as do all other mammals. Young monotremes therefore do not suckle, but lap up milk as it is secreted by the mammae. Female platypuses also differ from other mammals in that they have only one ovary, located on the left side of the body, whereas other mammals have two ovaries, one on each side.

The subclass Theria contains all other living mammals that give birth to live young. It is divided into two infraclasses: the Metatheria, which has only one order—the Marsupialia, or pouched mammals—and the Eutheria, or placental mammals. Female marsupials have two uteri (singular, *uterus*), the muscular organs in which young develop. Marsupials give birth to live young, but they are born blind and greatly undeveloped. As soon as they are born, they crawl through the mother's fur to her teats,

Land-dwelling migrating mammals have been greatly reduced in number because of human population of the globe. Mammals that still migrate include the caribou and reindeer, which inhabit the northern expanses of Canada and arctic Europe.

where they clamp on with their jaws and remain attached for several months. Most marsupials have pouches, in which the young are protected as they develop. However, many mouse-sized species lack pouches, and the young dangle exposed during their period of attachment. Kangaroos and other marsupials leave the pouch before they are weaned.

In contrast to marsupials, eutherian development takes place mostly within the mother and most young are born at an advanced stage of development. Eutherians have only one uterus, although in some it is divided into two halves. As with all mammals, fertilization is internal. Once the female's egg, or ovum, has been fertilized by the male's sperm, it becomes fixed to the wall of the uterus. As in birds and reptiles, special structures then develop around the embryo. These are the amniotic sac, or amnion, which contains amniotic fluid, and the allantois, or respiratory membrane. In addition, a placenta is produced from tissues that arise from both mother and embryo. The placenta is usually attached to the uterus wall and connected to the embryo by the umbilical cord. The purpose of the placenta is to allow the passage of oxygen and nutrients from mother to embryo and waste products from embryo to mother. These substances are carried in the blood and pass through a thin partition separating the two blood systems. This form of reproduction is highly efficient and allows the fetus to remain inside the mother until it is very well developed. Even so, in most mammals the newborn are not fully mobile and are often helpless.

The testes, or sperm-producing organs, of most male mammals lie in two sacs called scrota (singular, *scrotum*) that protrude on the underside of the body just in front of the pelvis and fuse behind the penis. In marsupials, however, the scrotal sacs lie in front of the penis. The testes lie outside the body because high body temperatures hinder the process of sperm maturation. In many mammals, such as some primates, the sacs descend during breeding season and are retracted when the breeding season is over.

The length of time young mammals remain physically, or at least nutritionally, dependent on the mother varies from group to group. Some newborns are able to fend for themselves when they are only a few days old. Others may not become physically independent for many months. In many species, such as

elephants, chimpanzees, and wolves, which live in family groups, the young may remain with the mother for many years. Elephants and other community animals cooperate in finding food, avoiding danger, and caring for the young. This life style necessitates some sort of code of behavior for herd members, which is learned from the parents and other members of the group.

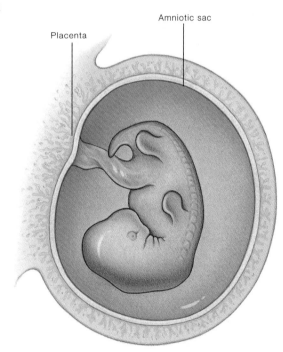

Placenta

Amniotic sac

A developing human embryo, like most mammals, receives nourishment from the mother through an organ called the placenta. The placenta attaches the embryo to the wall of the mother's uterus. An amniotic sac forms around the embryo and is filled with fluid. The fluid protects the embryo by absorbing jolts to the mother's body.

Monotremes and marsupials

The classic description of a mammal is an animal that gives birth to fully formed young that are initially nourished on mother's milk. But there are two groups of mammals that defy this definition: the monotremes and the marsupials.

The monotremes (subclass Prototheria) lay eggs, and the marsupials (subclass Metatheria) give birth to embryonic young that continue their development within the protection of the mother's abdominal pouch. The six species of monotremes and approximately 270 species of marsupials together make up only about 4 per cent of all mammal species.

The platypus inhabits the rivers of eastern Australia. This monotreme searches for food on the riverbeds, probing for small crustaceans and worms with its sensitive bill. Its body temperature is lower than that of other mammals at 86° F. (30° C) and is not constant, but fluctuates slightly with its environment. It has an average life span of 11 to 12 years.

The koala of eastern Australia feeds almost exclusively on eucalyptus leaves and bark. The female of this tree-dwelling species usually produces a single offspring every other year, which she carries in her pouch for six months and on her back for six months. Koalas measure from 25 to 30 inches (64 to 76 centimeters) in length and weigh 15 to 30 pounds (7 to 14 kilograms). On their front limb the first and second fingers are opposable with the other three—that is, the tips of the first and second can be placed opposite the other three. This aids them in grasping tree trunks and branches.

Monotremes

The monotremes are the platypus and the echidnas, or spiny anteaters. The most striking difference between monotremes and other mammals is that monotremes lay eggs covered by a leathery shell. However, like other mammals, monotremes feed their hatched young on milk from mammary glands.

Monotremes are curiously reptilian in other ways as well. Just as in reptiles, the monotreme excretory and genital ducts have a common opening known as the cloaca. It is this feature that gives monotremes their name, which means "one hole." The structures of the bones in the lower jaw and middle ear are similar to those of other mammals, but the girdle of bones that supports the forelimbs is reptilian in form. In addition, the brain and circulatory systems of monotremes are mammalian but have some reptilian features. This peculiar mixture of characteristics indicates that monotremes separated from the therian evolutionary lineage before many of the therian characteristics evolved.

There are two families of monotreme: Ornithorhynchidae (bird-noses), which contains a single species, the platypus, or duckbill, and Tachyglossidae, which is made up of two species of echidnas, also called spiny anteaters. The platypus, which is found only in Australia, is a streamlined animal with a flat snout like a duck's bill. It lives in burrows that it digs in riverbanks or inherits from other animals. It hunts for food underwater, depending largely on the tactile sense organs on the soft edge of its bill to find its prey, which includes small animals such as larvae, earthworms, and crustaceans.

The platypus has webbed feet and nostrils on the tip of its bill through which it breathes while it floats on the surface of the water. When it dives, a protective skin fold closes over its ears and eyes, and it can stay submerged for up to five minutes. Male platypuses have a venom spur on each hind leg, as do male spiny anteaters.

Spiny anteaters, or echidnas, occur in the sandy and rocky regions of Australia and New Guinea. They have a protective coat of short, sharp spines, rather like a hedgehog. These animals have no teeth but feed on termites, ants, and other insects by picking them up with their long, sticky tongue. The female spiny anteater temporarily develops a pouch on her abdomen during breeding season, into which she transfers her eggs after laying. The young hatch about a week later and feed from the milk ducts, which open into the pouch. They stay in the pouch for several weeks until their spines have developed.

Marsupials

Marsupials are now represented by only about 270 species, all of which live either in the Americas or in Australasia. However, the fossil records of the Eocene age, which began about 58 million years ago, indicate that marsupials were once widely distributed and even more common than eutherians. One reason for the marsupial decline may be that their mode of reproduction is poorly adapted to unpredictable food supplies.

The most distinguishing feature of the marsupials is the pouch, or marsupium, which contains the mammary glands. Although there are many species in which the marsupium is not present, it is this feature that gives marsupials their name. The embryos of some marsupials are supplied with nutrients from the wall of the uterus, whereas in others there is no placental connection between the uterus and the embryo. In all species of marsupials, the embryo remains in the uterus for a very short period and develops quickly.

At birth, the embryo is tiny. At most it may be 1.2 inches (3 centimeters) long and many are no longer than a grain of rice. Its forelimbs and nervous center are well developed, but the hind limbs are mere buds. It crawls, unaided, from its mother's birth canal to her pouch, where it latches onto a nipple that injects milk into the young animal's mouth. A baby kangaroo, or joey, remains in its mother's pouch for about six months, and will return to feed or seek refuge until it is about eight months old.

The kangaroos, the largest and perhaps the best known marsupials, have forward-opening pouches, but the form of pouch varies among species. For example, wombats and phalangers, including the koala, have pouches that open to the rear. The arrangement is an advantage for the young of digging animals, although it means the mother cannot clean the pouch.

The marsupial reproductive system differs from that of eutherians in various other ways. For example, the females have a double uterus and double vagina, and in many species the males have a forked penis with the testes in front of it.

Kinds of marsupials

The marsupials are an extremely diverse group, and they include tree-dwelling, fruit-eating, grazing, burrowing, insect-eating, and meat-eating types. There are six families of marsupials. The opossums (Didelphidae) and the rat opossums (Caenolestidae) are found only in the Americas. Most didelphids eat any plant and animal food. Caenolestids live in western South America and eat insects and other invertebrates. Carnivorous marsupials (Dasyuridae) include insect-eating species often called marsupial mice as well as the native cat and the fierce Tasmanian devil. The insect-eating bandicoots (Peramelidae) use their long, pointed noses to root in the soil for insects. Phalangers (Phalangeridae) eat mainly fruits, flowers, and nectar and nest in holes in trees. Kangaroos and wallabies (Macropodidae) feed mainly on grasses.

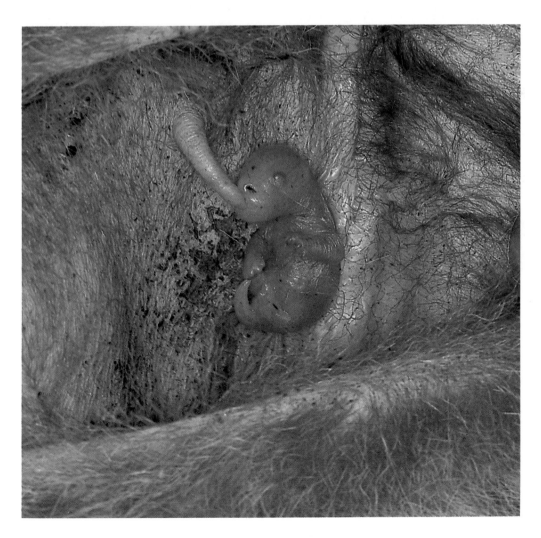

Young marsupials are known as neonates. In kangaroos, a single neonate emerges from the mother's cloaca about 33 days after conception. Its forelimbs are well developed, but its hind limbs are mere buds. It struggles through the mother's fur to her pouch, where it latches onto one of the teats. About six months later, the joey emerges from the pouch.

Insect-eating mammals

Insectivora is probably the most ancient living order of eutherian mammals. These insect-eating mammals are known to have lived as much as 100 million years ago, sharing their world with the dinosaurs. All insectivores share certain basic features, but each species is specialized for a different way of life. Along with such non-insectivore orders as bats (Chiroptera), which are generally considered to be descended from Insectivora; anteaters, sloths and armadillos (Edentata); pangolins (Pholidota); and some primates, insectivores form the majority of a large group of mainly insect-eating mammals. Although many species belonging to other orders, including rodents and foxes, also feed primarily on insects, this section discusses only those species whose entire family and order have evolved to consume insects and other invertebrates.

Insectivore classification

Today, Insectivora is the third largest order of mammals, occurring throughout the world and comprising four suborders and about 410 species. Most insectivores belong to the suborder Lipotyphla, which includes moles, solenodons, tenrecs, desmans, golden moles, hedgehogs, moonrats, and shrews. Other suborders include Zalambdodontia, or the water shrews (also known as otter shrews); Macroscelidea, or the elephant shrews; and Dermoptera, which contains the flying lemur, the most accomplished of gliding mammals.

Typically, insectivores are small, nocturnal mammals with elongated, narrow snouts. Most are distinguished from other small mammals by the fact that they have five digits on each limb—most rodents, for example, have four or fewer. Each digit has a distinct claw. The body is covered with short, dense fur. In some species, such as hedgehogs and tenrecs, some of the hairs are developed into spines. Insectivores often have small ears and small eyes with poor vision. Their senses of smell and hearing, however, are usually sharp. The brain has a primitive structure, and the placenta is simpler than that of most other mammals. Insectivores have up to 44 teeth—the largest number normally found in placental mammals. These usually have sharp cusps that enable them to slice their prey.

Insectivores include ground-living, burrowing, climbing, and even semiaquatic species.

They feed on insects, grubs, snails, and occasionally on helpless vertebrates, such as the young of ground-nesting birds. Many insectivores are extremely active and need to be refueled constantly by large quantities of food.

Habitats and life styles

The solenodon is about the size of a rat and has a pointed snout. It uses this sensitive snout to root about on the ground for invertebrates and plant material. It also eats lizards, frogs, and small birds. Like some other insectivores, it produces a toxic saliva from a gland in the lower jaw. Of the two species of solenodon, one lives in Cuba and the other in Haiti.

The tenrec family (Tenrecidae) includes about 20 species, all of which are found on Madagascar and the Comoros islands. They retain some reptilian features, such as a cloaca. Tenrecs have adapted to fill a wide range of habitats in Madagascar. Some resemble shrews. Those of the genus *Setifer* or *Echinops* have sharp spines and resemble hedgehogs. Others are similar to mice and moles. Their main diet consists of small invertebrates and plant material.

The family Talpidae consists of about 29 species of mole, shrew-mole, and desman, which occur widely in the Northern Hemisphere. Most are burrowers and are highly modified for an underground existence, with powerful spadelike front feet, tiny weak eyes,

The diversity of insectivores is reflected in their limb structures. All have basically pentadactyl (five-toed) limbs, modified to suit their life styles. Shrews walk on their toes, whereas a mole's digits form a broad spade for digging. Anteaters have an enlarged third digit with a curved claw for breaking open ant and termite nests. Bats have four elongated digits on the forelimbs, which support the membranous wing.

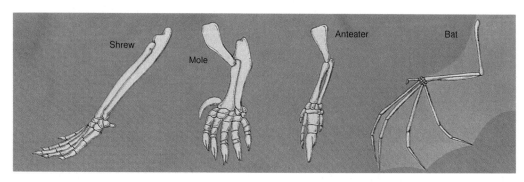

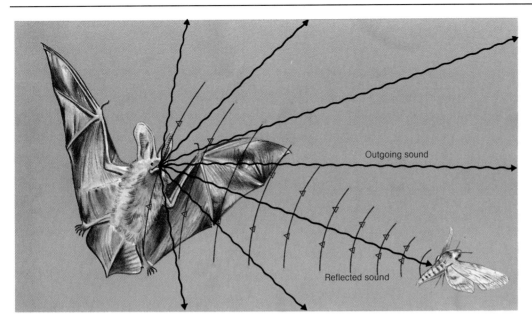

Most bats *(far left)* eat insects and hunt their prey at night using echolocation *(left)*. While flying through the air at up to 15 miles (24 kilometers) per hour, they generate high frequency sounds and detect prey or obstacles by the echoes they reflect. The range of detection is, however, limited to 3 feet (1 meter) or so, but is sufficiently discriminating to enable some bats to catch fish by detecting the ripples they make on the water surface.

Outgoing sound

Reflected sound

and no external ears. The short, fine hairs on the body can be brushed in any direction, enabling them to reverse easily in their tunnels. The desmans, which are the largest of this group, lead a semiaquatic life in or near ponds and streams, feeding on aquatic invertebrates and fish. They use their long noses as snorkels, turning them up so that they stick out of the water.

Golden moles, which resemble true moles, form the family Chrysochloridae and are found in a range of habitats in Africa. They have a blunt snout covered with a leathery pad. The eyes are tiny and weak, and the front feet are armored with powerful claws for efficient burrowing. Golden moles are diurnal—that is, they are active during the day—and feed on a range of underground invertebrates. Some desert-dwelling, sand-burrowing types, such as the desert golden mole, even capture and kill burrowing reptiles.

The hedgehog family (Erinaceidae) includes the familiar hedgehogs, with their covering of sharp, tough spines, and the more rodentlike, long-snouted moon rats. There are 17 species in all, found in Europe, Asia, and Africa. Erinaceids are nocturnal. They feed on invertebrates, carrion, and small mammals.

The shrews of the family Soricidae are the most successful insectivore group in terms of numbers. There are about 250 species, and they inhabit almost every corner of the world. Most shrews are small. The smallest, the pygmy shrew, weighs less than 0.09 ounce (2.5 grams). Shrews forage on the ground, moving under logs, plant debris, and into crevices in search of invertebrate prey. Some are able to subdue larger animals such as mice and frogs because, like solenodons, they also have a highly toxic substance in their saliva. The prey is rendered helpless by just a few bites from these venomous animals. Shrews communicate by high-pitched squeaks and high-frequency noises that humans cannot hear.

The water shrews of the order Zalambdodontia are distinguished from other insectivores by their five-cusped molars, with the cusps arranged in a W-shape. This formation is almost identical to that of the earliest eu-

therian mammals. Water shrews lead a semiaquatic life in streams, rivers, and swamps, feeding on fish, crabs, and frogs. The giant otter shrew is the largest living insectivore, with a length of about 24 inches (60 centimeters). With its sleek, dense coat and powerful tail, it resembles a small otter.

The elephant shrews differ from other insectivores in that their hind limbs are adapted for hopping. There are about 18 species of elephant shrews, and they are found in Africa in habitats ranging from semidesert to forest. Elephant shrews have large eyes and ears and are further distinguished by their extremely long, narrow snout. These animals forage on the ground using their mobile snouts and long-clawed front feet to search for small invertebrates.

Hedgehogs are mainly nocturnal animals and predominantly insectivorous, although they have been known to feed on carrion and small mammals. They hibernate in winter because it is difficult for them to find and digest enough food to combat the cold.

Flying lemurs

The two surviving species of flying lemur (also known as colugos) represent an early specialized development of the basic insectivore type. These Southeast Asian tree-dwelling mammals resemble true lemurs, but are not related to them. They also are not really able to fly, but they glide among the treetops by means of membranes at the sides of the body. These membranes extend from the neck to the front feet, the hind feet, and the tip of the tail. Normally held folded in at the sides, the membranes become parachutelike when the animal extends them by stretching out its limbs. Flying lemurs are excellent climbers, and because they are nocturnal, they spend the day

The three-toed sloth is found in the Amazonian forests, where it feeds on leaves and rarely descends to the ground. Its three digits are internal for almost their entire length. Only the long, curved claws protrude from the skin.

The tamandua, or collared anteater, is a mainly nocturnal, tree-dwelling species found in South America. Its body length reaches 23 inches (58 centimeters). This animal walks on the outside of its hands to prevent the tips of its curved claws from digging into its palms.

hanging by their clawed feet from a branch. At night they glide from tree to tree, feeding on leaves, buds, flowers, and fruit. Their jaw is unusual among mammals in that its lower incisors have become "comb teeth" and are used for grooming the fur. The comb teeth are similar to those of the lemurs.

Bats

The bats are second only to rodents in terms of number of species, more than 900 of which are known from almost all parts of the world. Their order, Chiroptera, is divided into two suborders. The suborder Megachiroptera contains one family, known as fruit bats, which feed primarily on fruit. Microchiroptera contains the other 17 families, most of which are largely insect-eating.

The main reason for the enormous ecological diversity of bats is their use of an otherwise unoccupied ecological niche—their ability to capture nocturnal insects in flight. No other mammal can actively fly, and no other creatures, even birds, can surpass the bats' mastery of the night skies. This agility is achieved by their sonar sensory system for navigation and prey-finding. To enable them to fly, bats have evolved extensive wings that consist of thin skin flaps stretched between the elongated fingers of the hands and between the forearm and the smaller hind legs. In most species, the skin flap also extends backward from the legs to the tail. The skin is an extension of the back and belly. In flight, bats can maneuver exceptionally well.

When bats alight, they hang upside down from perches with the aid of clawed hind feet. Some bats feed on the nectar and pollen of night-flowering plants and are important pollinators of those plants. Others feed primarily on fruit. For these species, flight enables them to travel long distances between fruiting or flowering trees. Other bats feed on small invertebrates and mammals and even capture fish in swoops to the water surface. The vampire bats (Desmodontidae) use their sharp, tri-

angular-shaped front teeth to cut a narrow groove into the skin of sleeping mammals and birds and to sever capillaries, which bleed freely. They lap up the blood that flows from these wounds, consuming about 1 tablespoon (15 milliliters) of blood a day.

Armadillos, anteaters, and sloths

The order Edentata today contains about 30 species in three families: armadillos (Dasypodidae), anteaters (Myrmecophagidae), and sloths (Bradypodidae). All have a reduced number of teeth, or none at all. And although the sloths are grouped with the insect-eating armadillos and anteaters, they are completely herbivorous—that is, they feed only on plants.

The 20 species of armadillo of the Americas are nocturnal and feed primarily on insects, which they lick up with their long, narrow tongue, as well as other small invertebrates and vertebrates and, in some cases, carrion. Their limbs are sturdy and powerful, with large strong claws. Because armadillos have only a few small teeth set well back in their mouths, they cannot bite in self-defense. Instead, their bodies are protected by a shell made up of bands of bony plates, leaving only the limbs and underside vulnerable to attackers. Two species of armadillo curl up into a ball to protect these parts of the body.

Anteaters, with their long, toothless snout and long, sticky tongue, are highly specialized for feeding on ants and termites. Four species of anteater live in Central America and South America. The largest of them, the giant anteater, grows to more than 6 feet (1.8 meters) long. The giant anteater is ground-dwelling, but the other species spend at least some of their time in the trees. The collared anteater, or tamandua, wraps its long tail around tree branches while feeding on ants and termites so that it will not fall.

The five species of sloth are so adapted to life in the trees that they can barely move on land. For this reason, they rarely descend to the ground, as they would make easy prey. They live in forests in Central and South America and spend much of their time hanging upside down from their large hooklike claws. There are two genera: *Bradypus*, the three-toed group, and *Choloepus*, which are two-toed. The three-toed species tend to be slower than the two-toed, and their grooved hairs carry algae, which give them a green color.

Pangolins

The pangolins (order Pholidota), or scaly anteaters, of Asia, Indonesia, and Africa lead a life similar to that of the American anteaters. They have no teeth and feed largely on termites, ants, and other insects, which they capture with delicate movements of their long tongue. The largest pangolin can extend its tongue 16 inches (40 centimeters) out of its mouth. Like anteaters, pangolins have a gizzardlike stomach that grinds down the hard exoskeletons of the insects they eat.

Pangolins are covered by a coat of horny scales resembling a coat of mail. When threatened, they roll themselves into a tight ball that few predators can pierce. Pangolins are shy creatures and move about at night.

Scaly anteaters, or pangolins, are distinguished by their scaly skin, which resembles the bracts of a pine cone. These arboreal mammals have a prehensile tail and a flexible body that can roll into a ball. Like the anteaters, they have strong claws for tearing open termite nests.

Primates

Human beings and their closest relatives in the animal kingdom belong to the order Primates. The name, which means "first ones," is apt because early members of this group lived at the same time as the dinosaurs, although they were not the earliest eutherian mammals. The primates began to separate from the other mammalian orders about 80 million years ago, at which time they differed from them very little in appearance and behavior. However, they bore the evolutionary potential whose expression since that time has produced the diverse forms of their descendants today.

The present order is divided into two suborders. The Prosimii include tarsiers, lemurs, indris, aye-ayes, galagos, pottos, and lorises. Some zoologists classify the tree shrews as primates. The Anthropoidea consist of the New World monkeys (tamarins, marmosets, and cebid monkeys), the Old World monkeys (macaques, baboons, and colobus monkeys), the lesser apes (gibbons and siamangs), and the great apes (orangutans, chimpanzees, and gorillas). Among the small primates is the pygmy marmoset, whose body length measures about 5.5 inches (14 centimeters). The largest primate is the gorilla, which stands about 6 feet (1.8 meters) high and weighs about 450 pounds (204 kilograms).

Distribution

The prosimian group is dominated by the lemurs and their relatives, which are concentrated in Madagascar and the nearby Comoros Islands. About 50 million years ago, what is now the island of Madagascar separated from mainland Africa and lemurlike prosimians on Madagascar were isolated. There were few other types of mammals on Madagascar, and lemurs were able to diversify with little competition. Lemurs that remained on the mainland developed a nocturnal lifestyle in response to competition from the more versatile Old World monkeys and apes. The nocturnal prosimians of the mainland include the galagos—sometimes called bush babies—of tropical Africa and the lorises (Lorisidae) and tarsiers (Tarsiidae) of tropical Asia. Prosimians failed to reach Australia because it was isolated in the Pacific well before their evolution. Tarsiers and lemurs once existed in North America but are now extinct on that continent.

The success of the primates meant that some species increased in size and moved from the ancestral forest habitats. Today the largest primates spend little time in the trees. Primates, whatever their way of life, have remained essentially animals of the tropics,

Lemurs, such as the ring-tailed species, inhabit Madagascar. They were once revered by the islanders as embodying the spirits of the dead. Since Christianity replaced these traditional spiritual beliefs, however, lemurs lost their special status and have been killed for food.

The primates consist of two suborders: the Prosimians, which includes tree shrews and lemurs, and the Anthropoids, which contains New World monkeys, Old World monkeys, and hominoids.

Prosimians	New World Monkeys	Old World Monkeys	Hominoids
Tree shrew (*Tupaia* sp.)	Marmoset (*Callithrix* sp.)	Mandrill (*Mandrillus* sp.)	Gibbon (*Hylobates* sp.)
Lemur (*Lemur* sp.)	Spider monkey (*Ateles geoffroyi*)	Colobus monkey (*Colobus* sp.)	Gorilla (*Gorilla* sp.)

where there is an abundance of leafy vegetation and large fruits for them to eat.

Evolutionary trends

The earliest primates are thought to have exploited an unoccupied environmental niche as fruit- and insect-eaters in the trees. This tree-dwelling tendency, and the adaptations that have resulted from it, have remained with the primates throughout their evolutionary history. The more distinctive developments among the primates include the evolution of hands and feet equipped with grasping fingers and toes. This involved the separation of the digits and the development of musculature that enabled them to fold their hands around branches or other objects. The first digit (thumb or big toe) became widely separated from the others but could be folded across the palm of the hand, providing a manipulative grasp. The tips of the digits became flattened, and in many species nails rather than claws developed on their dorsal surfaces. Friction pads of deeply folded skin developed to maintain a hold on smooth or slippery surfaces.

Another important development was the reduction in the size of the snout, which allowed the eyes to move forward and stereoscopic (depth-perceiving) vision to evolve. This was particularly important for judging distances when leaping among the trees. High activity levels and manipulative skills require enlarged brain centers, and the primates developed large brains in comparison with their body size. In general, larger brain-to-body ratio indicates a higher level of intelligence in animals, but this is not always the case. In addition, the development of the embryo in the womb is more complex in primates, and gestation is often longer than for other eutherian mammals of similar size. Parental care is also longer in most primates.

Most primates are highly social animals, whose life is characterized by complex interactions among members of a group. With so-

cial safety and prolonged care, education of the young can develop. Strong links are forged between an infant and its mother in the early stages of life. Later, bonds are made with playmates and other members of the group. Such developmental trends have progressed to different points in different primates but they are generally less advanced in the prosimians than in the anthropoids.

Prosimians

Zoologists disagree on the classification of the predominantly ground-dwelling tree shrews of the family Tupaiidae. Some zoologists consider them a primitive member of the primate order. Others place them in the order Insectivora, and still others classify them in an order of their own, Scandentia.

Zoologists agree, however, that tree shrews are related to both primates and insectivores and combine characteristics of both. Tree shrews have claws on each of their toes, unlike primates, which have at least one nail on each foot. Like insectivores, the teeth of tree shrews are adapted to a diet of insects, fruit,

The slow loris is, as its name implies, a slow-motion climber. The opposable thumbs and big toes of this animal give it a powerful grip as it pulls itself along, hand over hand, suspended under branches. It eats fruit, birds' eggs, and insects, which it stalks stealthily.

The prototype primates began to separate from the other mammalian orders about 80 million years ago. Today, the order contains about 230 species.

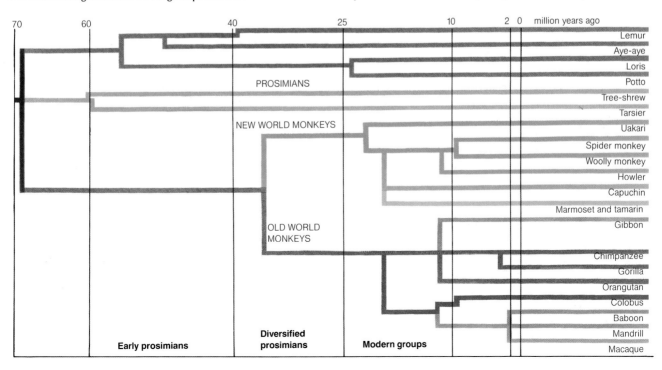

70   60   40   25   10   2 0   million years ago

PROSIMIANS

NEW WORLD MONKEYS

OLD WORLD MONKEYS

Lemur
Aye-aye
Loris
Potto
Tree-shrew
Tarsier
Uakari
Spider monkey
Woolly monkey
Howler
Capuchin
Marmoset and tamarin
Gibbon
Chimpanzee
Gorilla
Orangutan
Colobus
Baboon
Mandrill
Macaque

Early prosimians   **Diversified prosimians**   **Modern groups**

The digits of the tree shrews have long, sharp claws and poor grasping power. The thumbs are not opposable with the fingers even though they are longer. The soles of the feet are naked and have thick friction pads.

Foot   Hand   Grip

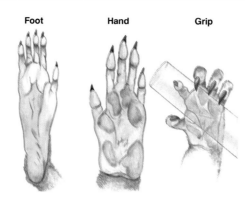

The African potto has short digits with nails, except on the second toe. The index finger is a mere stub and has no nails or joints. The digits lie almost completely opposite the thumbs and big toes.

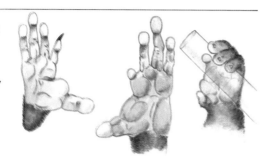

The tarsier's long digits with flattened tips and large friction pads contribute to its superior grasping power. It also has flattened nails rather than claws, except on the second and third toes.

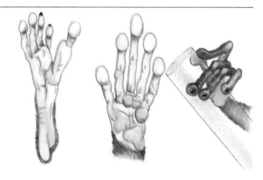

Most brachiating monkeys and apes have short thumbs and long fingers. A gibbon's thumb is longer, but is folded out of the way across the palm when the animal uses its arms to swing through the trees.

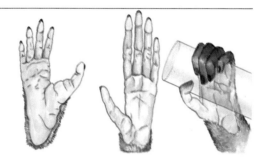

The hands and feet of macaques grasp well, having opposable thumbs and big toes with flattened nails. In the larger, less active primates, the friction pads provide large areas of skin for tactile nerve endings.

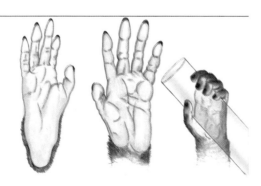

and worms. While primates have forward-facing eyes, the eyes of tree shrews are situated on each side of the head, and as a result they lack stereoscopic vision. However, their moderately large eyes and bone-encased eye sockets resemble those of primates. Like primates, tree shrews also have bony ridges called postorbital bars surrounding the eye sockets, rounded ears with folds, and relatively large brains.

Tree shrews inhabit the forests of tropical eastern and southeastern Asia. The name of their genus, *Tupaia,* is derived from the Malay word for a squirrel, which tree shrews resemble in size and, to some extent, in appearance. Tree shrews are small, often bushy-tailed creatures. They are nocturnal, and their eyes are highly sensitive to light but not to color. They have prominent snouts and depend greatly on their good sense of smell. Tree shrews generally produce twins. Each of the twins develops in one of the two long branches of the bicornuate (two-pronged) uterus common to the primitive placental mammals. Gestation lasts about six weeks and parental care is minimal.

The status of the tarsiers as primates is certainly less controversial than that of the tree shrews. These arboreal animals inhabit the forests of the East Indies and the Philippines, and one of their most notable features is the greatly flattened tips of their digits. These digits enable the tarsiers to jump and support themselves even on smooth, vertical surfaces. The ability to grasp and climb among smooth branches, which is where most trees produce their most succulent buds and leaves, enabled the early primates to live in parts of the forest in which no other mammals lived.

Travel between high fruit-bearing branches was made possible by the development of leaping as a means of locomotion. The tarsier gets its name from its means of leaping—its greatly elongated ankle bone, or tarsal bone, allows it to jump at least 6 feet (2 meters).

The tarsier has a relatively flat face and forward-facing eyes with stereoscopic vision, which aids it when jumping. Its enormous, fixed, bulging eyes are its most remarkable feature. Unable to move them, it can swivel its head a full 180 degrees to the left or right. The size of its eyes may contribute to its good night vision, but a more important factor is the abundance of rods in the retina, which operate well at low light intensities, and absence of color-sensitive cones. This arrangement is typical in nocturnal vertebrates because color vision and fine resolving power are less useful in the dark. Its eyes lie in eye sockets completely encased by bony ridges.

The largest number of prosimians are found among the lemurs. These monkeylike, tree-dwelling animals have forward-facing eyes with less well developed binocular vision than other primates. However, they have better night vision as a result of a reflector at the back of the retina, called a tapetum. With the exception of the second toe, lemurs resemble tarsiers in that they have nails rather than claws on their digits, which allow them to grip more easily. Unlike the tarsier, however, the thumbs of lemurs are somewhat opposable in that they can be used to grasp objects. Their nose is foxlike in shape, although not as pointed as that of the tree shrews. Scent, an

important feature in the life of these animals, is secreted from glands to convey a complex language of signals. It may be secreted to convey aggression or to mark territory, for example. The lower incisor teeth are modified to form a comb, which is used for cleaning the fur, as is the claw on the second toe.

Related to the true lemurs and also found in Madagascar are the indris and the aye-aye. The aye-aye was once thought to be a rodent, mainly because its curved, chisellike incisors continue to grow throughout its life. But if they did not, the teeth would be worn away because the aye-aye tears at rotten wood and gnaws at tree branches for hidden grubs detected by their sensitive ears. It then dislodges the prey with its extraordinarily long, thin, hooked middle finger. The finger is twice as long as the others and has a sharp claw.

In Africa, the prosimians are mainly represented by the galagos. These animals are fast-moving, with tremendous jumping power, and are nocturnally active on the ground or in the shrub layer of the forest. Their eyes are large and forward pointing, and the snout is short, giving them good binocular vision. They have the curious habit of urinating on their hands and feet, which probably helps them spread their scent for purposes of territorial claim. Galagos are unlike the lorises and pottos, which although similar in appearance, travel differently. They exert a viselike grip on a branch with their hands and feet while they move slowly upside down along the branch.

New World monkeys

Most of the primate group inhabiting the New World are considered to be more primitive than the other anthropoids. They include two families—the Callitrichidae, which consists of the more than 20 marmosets and tamarins, and the Cebidae, which contains 26 species, including howlers and capuchins. They are con-

fined to tropical forest areas and are, in general, diurnal animals. Only the douroucouli, or owl monkey, is nocturnal.

New World monkeys are termed platyrhine, because they have flat noses with widely spaced nostrils. They have better stereoscopic vision than the prosimians and also are more dexterous. These two complementary developments may partly explain why they displaced prosimians in the New World.

The callitrichids are the smallest of the New World monkeys, measuring 10 inches (25 centimeters) or less. They have curved, clawlike nails on all their digits except for the big toe, which has a flattened nail. These claws help them grip branches as they travel along on all fours. They cannot make use of the thumb to grasp and so they grasp objects between the fingers and the palm of the hand. Like the tree shrews, callitrichids usually have twins that are carried by their father, who hands them over to the mother for suckling.

Cebids have better manipulation than the callitrichids. Larger and heavier than the marmosets and tamarins, they have to hold onto branches rather than merely balance on them. Like the callitrichids, some cebid monkeys live in family groups. Others, however, are found in large, multi-male troops. The females generally tend the young entirely, but in one or two species, the males carry and care for them.

The prehensile tail

In addition to their manual dexterity, some of the South American monkeys are further supported by a prehensile tail. Only the howlers, the spider monkeys, woolly monkeys, and the woolly spider monkey have a prehensile tail. Howler and woolly monkeys use their tail as an extra hand. The tail is most fully developed in the spider monkey, which can use it to support its entire weight. The end of the underside of the tail is naked and has ridged skin,

The tiny marmosets and tamarins of the New World often have large tufts of fur that conceal their ears, such as this golden lion tamarin. Marmosets and tamarins typically live in small family groups that include an adult male, an adult female, and offspring up to one or two years of age. Marmosets are often found feeding on fruit and invertebrates in the middle to upper levels of the forest canopy.

which provides a better grip. It is also very sensitive and can pick up even small objects.

New World monkey anatomy and senses

The spider monkeys are typical of New World monkeys in that they move through the trees by jumping from branch to branch as well as swinging by their arms. This latter method of locomotion is called brachiation, the supreme exponents of which are the Old World gibbons. Like many brachiators, spider monkeys have hooklike hands with small thumbs, and arms that are much longer than the legs.

The sense of smell is important to these monkeys. Marmosets and tamarins mark their

Prehensile tails are particular to only a few New World monkeys, such as this howler. Howlers are the largest of the New World monkeys, having a body length of about 24 inches (161 centimeters) and weighing 12 to 20 pounds (15.4 to 9 kilograms). Because of their weight, they can use their tails only as an extra hand. The tail of the lighter spider monkeys, on the other hand, can support their whole body weight.

territory and points of reference with scent released from glands in the scrotal and anal regions. Capuchins rub their chest glands on branches for the same purpose and also urinate into a hand and spread the urine onto a foot before moving along their territory.

Color vision and visual acuity are also significant to the New World monkeys, with the exception of the nocturnal douroucouli. The bald uakaris, for example, have brilliant red faces set off by their white or red-brown fur. As well as species recognition, color also plays a role in food recognition—those with color vision can evaluate the ripeness of fruit or the freshness of foliage by looking at it.

The brains of the New World monkeys are much larger and more complex than those of the prosimians. The greater size of the brain is largely attributable to the expansion of the cortex, which coordinates sensory and motor functions and controls memory and intelligence. This overall trend in the dominance of the cortex (corticalization) is more advanced in the cebid than in the callitrichid monkeys.

Males are, in most cases, larger than the females. Like most other animals, they are territorial. The male howlers are most notable for the method in which they proclaim their dominance over their territory. They have an enormous hyoid bone that makes a cup-shaped voice box in the throat. This air sac is inflated to resonate sounds that carry a distance of about 2 miles (3.2 kilometers).

Few New World monkeys show evidence of menstrual cycles. Actual menstruation is rare and minimal, and swelling of the female external genitals is minor. Gestation lasts about 18 to 32 weeks, and their infantile, juvenile, and adult phases are longer than those of the prosimians. Prolonged parental care is a trend further developed in the higher primates as it is critical to learning and to the evolution of complex, flexible social behavior.

Old World monkeys

Old World monkeys are found in Africa and

Coloration is important in displays among many monkeys. The red uakari is most notable for its bright red face and almost bald head. The uakaris are the only New World monkeys with short tails. Troops of uakaris inhabit the upper layer of the Amazonian forest and rarely descend to the forest floor.

the warmer parts of Asia. They live in a wide range of habitats and are far more varied in their ways of life than are the cebids or marmosets. Some Old World monkeys inhabit semiarid areas beyond forest boundaries, where they are forced to hunt on the ground for any food they can find. These monkeys are all grouped in one family, the Cercopithecidae, which is divided into two subfamilies: the Cercopithecinae, which includes the macaques, the African baboons, and the guenons; and the Colobinae, which includes the langurs, the Colobus monkeys, and the proboscis monkey. The subfamily divisions reflect different eating habits. Most colobine monkeys are leaf-eaters, whereas the cercopithecine monkeys, most of which are ground-dwelling, generally eat whatever foods are available. Most species of cercopithecine monkeys have cheek pouches, in which they may carry food that they do not eat immediately.

The most obvious difference between Old World and New World monkeys is that the nostrils of the former are closer together and point downward rather than sideways. They are therefore known as narrow-nosed, or catarrhine, monkeys. Another distinguishing feature of all catarrhine monkeys is the presence of bare patches of hardened skin on the rump, called ischial callosities. These patches have neither nerves nor a blood supply and allow the monkeys to remain in a sitting position for long periods without serious discomfort.

Monkey limbs, locomotion, and hands

Monkeys usually walk and run on all fours, either on tree branches or on the ground, but many can stand and even run on their legs for short periods. They usually stand or run on their hind legs when carrying food, peering over high grass, or threatening enemies or other members of their group. The legs of most species are slightly longer than the arms.

The wooded savannas and open dry grasslands of sub-Saharan Africa are the home of primates that tend to move on all fours. The patas monkeys have exceptionally long limbs

relative to their trunk, enabling them to sprint through the grass with long, bounding strides, faster than any other primate.

Many Old World monkeys have fine manipulative skills that are associated with true opposability of the thumb—that is, its tip can be placed opposite any of the other fingers on the same hand. True opposability depends on the movement of the carpo-metacarpal joint and development of the thenar muscles at the base of the thumb. New World monkeys, in contrast, cannot move their thumbs at the base, but only at the second joint, or metacarpo-phalangeal joint. Thus, they do not have true opposability. The predominantly

Tree shrew

Lemur

Squirrel monkey

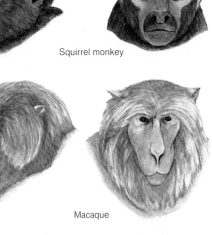

Macaque

Gorilla

One of the first adaptations of the early primates to an arboreal life was the reduction of the long snout found in the tree shrews. This reduction enabled the eyes, which are on the sides of the *Tupaia*'s head, to move closer together and achieve the binocular vision that is essential for accurately judging distances. This development led to improved abilities for leaping and brachiation. Lemurs are more arboreal than tree shrews, but still spend time on the ground. Their foxlike muzzle is proportionately shorter, and their eyes are closer together than those of the tree shrew. However, lemur binocular vision is not as well developed as that of the squirrel monkey. As with all New World monkeys, squirrel monkeys are platyrrhines, with flattened faces and sideways-facing, wide-set nostrils. In contrast, the nostrils of catarrhine Old World monkeys, such as the macaques, are close together and point downward. The great apes, such as the gorilla, have excellent binocular, or stereoscopic, vision.

Limb lengths of primates vary. The arms of woolly spider monkeys and gibbons, which brachiate, are longer than their legs. Gibbons' arms are especially long, and they rely solely on them for support, whereas spider monkeys also use their prehensile tail. The arms of gorillas, which live chiefly on the ground, are only slightly longer than their legs.

Monkey anatomy and behavior

The males and females of cercopithecine monkeys are very different in appearance—a phenomenon called sexual dimorphism. A female, for example, is only one-third to one-half the weight of a male.

Male physical prowess is important both inside and outside the group in determining the social pecking order and in defense of the troop. Many open-country monkeys are at risk from predators, and their social organization helps them to minimize the danger.

The powerfully built male baboons, geladas, and mandrills possess fearsome canine teeth, whose display, rather than actual use, is often enough to discipline troop members or to frighten off predators or strangers. Their teeth are housed in prognathous jaws—that is, jaws that extend beyond the upper part of the face—but the monkeys still have good binocular vision.

Higher primates have a well developed visual center in their brains. This is necessary because they communicate by color. Mandrills have brilliant blue and red facial and genital markings. The colors become even more intense to express arousal or excitement.

Monkey reproduction

In both Old World and New World monkeys, female reproductive behavior is seasonal. Many cercopithecine females, however—particularly mandrills, some of the macaques, and colobus monkeys—advertise their peak of fertility to the males by pink, swollen genitalia. Old World primates also menstruate more heavily than the New World primates and have regular menstrual cycles. Gestation and the postnatal phases are longer in the Old World monkeys than in the New World monkeys.

The apes

The last group of the primates, the superfamily Hominoidea, includes the great apes. The

ground-dwelling baboons and mandrills have relatively long, highly opposable thumbs and use them to pull up grass and other plants. Colobus monkeys, on the other hand, grip between their fingers and the palms of their hands because, like the spider monkeys of South America, their thumbs are very short or entirely absent. This feature is the source of their name, which means "maimed." Also like the spider monkeys, colobus monkeys brachiate (swing by their arms) through the trees.

The feet of most monkeys are larger and more powerful than their hands. All monkeys have five toes on each foot. The big toes look and function much like thumbs, giving the monkey an extra pair of grasping "hands."

Macaques are the most widespread of monkeys. The Japanese macaques are some of the few Old World monkeys found in areas where heavy frosts and snow are frequent. They have adapted by growing long beards and thick fur. These animals show remarkable ingenuity and exercise of choice in their eating habits. This mother and her infant are washing sweet potatoes in seawater because they prefer salted potatoes. They are also known to throw rice grains into the water to wash away the husks.

main characteristic that distinguishes apes from monkeys is their lack of an external tail, although as with humans, there are small, internal tail bones. Like the brachiating monkeys, those apes that brachiate have very long arms, which are longer than their legs.

The gibbons and the siamangs of Asia are often referred to as the lesser apes because they are smaller than the others. The gibbons populate the tropical forests of southeastern Asia, Malaysia, and Indonesia, whereas the siamang is confined to those of Sumatra and mainland Malaysia. The other Asian ape is the orangutan of Borneo and Sumatra. Together with the gorilla, the chimpanzee, and the pygmy chimpanzee of the African tropical forests, they constitute the great apes.

Ape locomotion and anatomy

The Asian apes are the most accomplished brachiators of all the primates. The siamang, which may weigh up to about 28 pounds (13 kilograms), and the still lighter gibbons are tremendously agile in the trees and rely almost exclusively on brachiation to move about. Effective brachiation requires the smooth rhythmical transfer of the body weight from one arm to the other, as the hands alternately curl around and release a branch. But this method of locomotion is not without its dangers. Many animals fall at some time during their life, and many skeletons of brachiating apes that have been studied show evidence of broken bones. Unlike the spider and colobus monkeys, the brachiating apes have retained their thumbs, though they are short in relation to their fingers and the rest of the hand. During brachiation, the hand assumes the shape and function of a hook. The thumb plays no role at all and is folded across the palm. On the ground, gibbons and siamangs can run and walk bipedally for short distances. Because their arms are so long, they hold them out at their sides for balance and to avoid dragging them on the ground.

In contrast to the lesser apes, the great apes do not rely heavily on brachiation as a mode of transport. The orangutan walks on branches while holding on to branches above with its hands. Gorillas and chimpanzees walk on the knuckles of their hands and on the curled toes of their feet. Gorillas have better opposability and make greater manipulative use of their thumbs than do the other apes, but it is the chimpanzees that make and use tools.

Whereas monkeys use color for purposes

of display and threat, the apes use their size. In addition, the orangutan uses the fatty deposits around its face, called "blinkers," for display. Blinkers contain reserves of fat that can be drawn upon in times of food shortage.

Vertebral column and skull of apes

In contrast to monkeys, which carry their slender bodies horizontally on four limbs, apes walk erect or semiupright. Apes also have fewer vertebrae in their trunks, which are wider than they are deep, whereas monkeys' chests are deeper front to back than they are across. Apes thus demonstrate the evolutionary trend toward the reduction of trunk length in the more recently evolved primates. Instead of a flattening of the face, the great apes have developed prognathic (protruding) jaws. Such heavy jaws and high skull capacity are associated with increased brain size and result in a heavy skull. The forward position and the weight of the skull make the head fall toward the chest. This feature and an enlarged crest, which runs from front to back across the top of the head, is characteristic of many great apes. Males twitch their

Long fringes of white fur are the dominant feature of the black-and-white colobus monkeys of Africa. This beautiful coat is the reason that they are hunted for the fur trade. These forest-dwellers are leaf-eaters and have a complex digestive system to digest tough foliage.

Brachiation, the method of movement under branches by swinging from hand to hand, has been perfected by the gibbons. The males also use this swinging and dropping action as a territorial display to neighbors.

crest in displays of aggression, but its height also serves to attract females.

Ape reproduction

The female chimpanzee alone among the apes exhibits external genital swelling as an indication of estrus. Female ape reproductive behavior is characterized by light menstrual bleed-

Gorillas are highly social animals and live in large troops of up to 35 animals. Each group is ruled by one large silver-backed male. Their life is one of complex interactions among the members of the group, the rules of which are taught at an early age and upheld by the dominant males. Of the three gorilla subspecies, two live in the lowland African rain forests of west and central Africa, and one, the endangered mountain gorilla, lives in the cool, evergreen cloud-forests of central Africa.

ing, and the sexual cycles last about a month. Gestation lasts about 30 weeks in gibbons—compared to 27 weeks in the much larger baboons—and up to 36 weeks in gorillas and orangutans.

Homo sapiens

Evidence from fossils has convinced most scientists that human beings evolved over millions of years from ancestors that were not completely human. The fossil record does not, however, provide enough information to trace human evolution in detail. As a result, not all experts agree on how humans evolved. Scientists, however, classify human beings as primates, and a survey of the living primates reveals that humans are most closely related to the apes and resemble them in a number of ways. Those trends that are preeminent in the evolution of the primates, such as the development of grasping hands, good stereoscopic eyesight, enhanced development of the cerebral cortex and other parts of the brain, and longevity, are far advanced in the modern apes but have progressed furthest in human beings. Humans, however, are vastly different from the apes in that they walk on two feet rather than on their hands and their brain is more than twice the size of that of any other living primate.

Hominid ancestors

Hominids are members of the group of species that includes human beings and our close prehuman ancestors. Hominid fossils, recovered from paleontological sites throughout the world, have been classified into several genera. New finds are increasingly frequent,

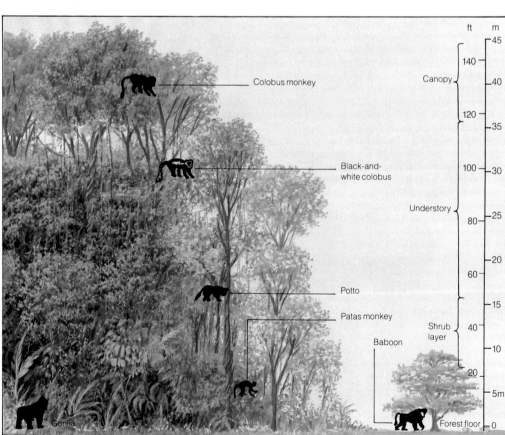

In an African tropical forest, the layers occupied by different primate species are separate. Boundaries are observed carefully to avoid confrontations and to reduce competition for the same food. The red colobus feeds in the upper layer of the forest, or canopy, which reaches to about 150 feet (45 meters) above the ground. The black-and-white colobus, however, feeds in the understory at heights of about 100 feet (30 meters). Pottos and galagos eat among the upper branches of the understory, whereas the patas monkeys feed in the lower trees and on the ground. Gorillas inhabit the floor of the denser parts of the forest and eat ground vegetation, as do the baboons, which live on the open savanna.

and scientists are constantly revising their understanding of human evolution.

The earliest known species of hominid, the australopithecines of southern and eastern Africa, lived from more than 4 million to about 1 million years ago. Australopithecus was slender and about the same size as the modern male chimpanzee with a similar cranial capacity of 24 cubic inches (400 cubic centimeters). In australopithecines, the hole in the skull through which the spinal cord runs up to the brain (the foramen magnum) is relatively far forward and vertical. This suggests that the spinal cord must have entered the brain perpendicularly and the skull on the vertebral column would have faced forward rather than slightly downward, in contrast to the great apes. The australopithecine pelvis is bowl-shaped with short, stout hipbones that extend backward; their hip surface area permitted the attachment of powerful gluteal (buttock) muscles, and their pelvic bones allowed the anchoring of strong abdominal and back muscles. In these and other respects, many of their skeletal elements are similar to those of modern humans. The australopithecines were undoubtedly bipedal and walked upright.

Other hominids shared parts of Africa with the australopithecines. But whereas the australopithecines may have used very simple tools—usually stone fragments, which they found but did not make—*Homo habilis*, or "handy human being," definitely made them. With a large cranial capacity of about 30 to 50 cubic inches (500 to 800 cubic centimeters) and less massive jaws, *Homo habilis* resembled modern humans more than did the australopithecines.

Homo erectus, the immediate predecessor of modern humans, emerged about 1.5 million years ago. These early bipedal hominids were neither very large nor particularly strong, nor could they run fast. It is very likely then that cooperation based on communication and sign language in food-gathering and group defense would have been vital for the survival of *Homo erectus*.

Except for *Homo sapiens*, or the modern human, *Homo erectus* had the largest brain of all the primates, accommodated in a cranium with a capacity of 42 to 76 cubic inches (700 to 1,250 cubic centimeters). Despite the evolutionary tendency within the primates for the skull to become domelike, *H. erectus* still had a sloping forehead with heavy, bony brow ridges and jutting jaws without a chin. But despite its more primitive appearance, the brain of *H. erectus* was complex and efficient.

H. erectus ushered in Acheulian stone technology, which featured a wide range of carefully manufactured tools for the killing or butchering of game and the preparation of food. This species was perhaps a good hunter. Paleontologists believe that deformations observed in fossil *H. erectus* bones were caused by excessive vitamin A intake as a result of eating too much raw meat. Fossil remains of these hominids in East Africa, Europe, China, and Java point to their northward and eastward expansion out of their native Africa.

Modern *Homo sapiens*

According to scientific theory, the modern human being, or *Homo sapiens*, became a recognizable species about 450,000 to 100,000 years ago. Earlier fossil remains of *H. sapiens* have been found across Europe into Asia. They gradually spread from western and eastern Eurasia to America and Australia, walking over land bridges created by lower water levels during ice ages. Human beings, at this stage, basically resembled *H. erectus*, but had a larger brain and smaller jaws and teeth. As time passed, *H. sapiens* began to look like today's human beings.

Although their facial muscles enable human beings to form a great many expressions, humans nevertheless rely primarily on symbolic spoken language to express and communicate most aspects of its culture. Of all animals, this characteristic is unique to human beings. Certain animals, including apes and monkeys, communicate by making a wide variety of sounds. Although these sounds express emotion and may communicate simple messages, they apparently do not symbolize any object or idea, as does the language of humans. Language therefore distinguishes human culture from all forms of animal culture.

Bones of *Homo erectus* were found in 1974 on the western shore of Lake Turkana in Kenya, Africa. Discovered between two layers of volcanic ash, the remains were dated by scientists as being about 1.5 million years old, making them among the oldest *Homo erectus* finds. Because the skeletal remains are so nearly complete, have not been chewed by scavenging animals, and display no evidence of disease, scientists feel that future study of this find will provide valuable insight into the anatomy, growth, and development of these early hominids.

Rodents and lagomorphs

The order Rodentia is the largest mammalian order and is divided into three suborders: Sciuromorpha, which contains the squirrellike rodents; Myomorpha, or mouselike rodents; and Hystricomorpha, or porcupinelike rodents. The word *rodent* means "gnawing animal." These mammals were perhaps so named because their large incisors and their manner of eating are their most obvious characteristics.

Lagomorphs were once thought to constitute a suborder of Rodentia because of their long, large incisors, but they are now treated as a separate order, called Lagomorpha. The order contains two families: the Leporidae, the hares and rabbits, of which there are some 50 species, and the Ochotonidae, the pikas, or mousehares, with 14 species. Rodents and lagomorphs are similar in appearance and habits, but differ in certain aspects of anatomy. They are all relatively small animals and are found in most parts or the world.

General features of rodents

The two long pairs of chisel-shaped incisors in each jaw, characteristic of all rodents, project from the mouth and are used to gnaw on hard foods, such as nuts and wood. These teeth, which are segments of a circle, grow continuously and must constantly be worn down at the tips. Rodents have no canines. Because these animals feed on tough materials that are difficult to break down, they have a highly developed cecum—a branch of the gut at the junction of the small and large intestines—that aids digestion. Living in the stomach are bacteria that break down cellulose, which then can be absorbed into the rodent's body.

The limbs of rodents are constructed for plantigrade locomotion. There are usually five digits on the forelimbs and three to five on the hind limbs. The digits on the forelimbs are used for holding food. Hind limbs may be adapted for running, jumping, climbing, or swimming (in which case the three to five toes are webbed). The females are capable of bearing numerous young at one time. Some species can reproduce almost without interruption throughout the year. The uterus is usually divided, although in some species it is double. Most newborn rodents emerge both blind and naked and are entirely dependent on the mother for several days. But the guinea pigs (Caviidae); spiny mice; and nutria, or coypu, have precocious young that are active at birth and can soon take care of themselves.

Squirrellike rodents

Squirrels, marmots, gophers, and beavers make up the suborder Sciuromorpha, in which there are seven families. These animals are found everywhere except in Australia, Asia, Madagascar, and parts of South America. They have a variety of habitats. Some, such as the squirrels (Sciuridae), live in trees. Others, such as gophers, burrow underground. Still others, such as beavers (Castoridae), live a semiaquatic existence.

The squirrels are particularly noticeable for their bushy tails and large eyes and ears. They are agile and graceful, and they move equally well on the ground and in trees. The ground squirrels, such as prairie dogs, generally make complex underground tunnels and chambers, some of which extend for hundreds of feet. Ground squirrels sometimes use the nests or burrows for storing food, which they eat when food is scarce. The marmots and woodchucks are also ground-dwelling species of squirrels. Like the prairie dogs, they live in large social groups. They hibernate during the winter in their underground chambers and live off reserves of fat built up during summer feeding.

Beavers are the only members of the family Castoridae and are distributed throughout the Northern Hemisphere. Primarily water-dwelling rodents, they have a dense under-

Rodent incisors grow continuously from roots that extend deep into the jawbone. The fronts of the incisors have only a hard coating of enamel, which acts like a blade as the dentin behind it is worn down. The surfaces of the cheek teeth have ridges that help grind the plant material that most rodents eat. Rodent teeth are ever growing and require hard or abrasive food, or gnawing on hard surfaces, to keep them at their proper length. If the teeth are prevented from wearing—that is, if the animal only eats soft food—the tips of the top and bottom teeth may grow past each other, and those of the lower jaw may perforate the palate or grow out of the mouth. The jaws that work the incisors and cheek teeth are controlled by masseter and temporal muscles. The terms sciuromorph (squirrellike), histricomorph (porcupinelike), and myomorph (mouse-shaped) are often used to categorize how the masseter muscles are arranged. Scientists generally believe that the histricomorphs are the most powerful jaw muscles, the sciuromorphs the most flexible, and that the myomorphs may have the strongest grinding action.

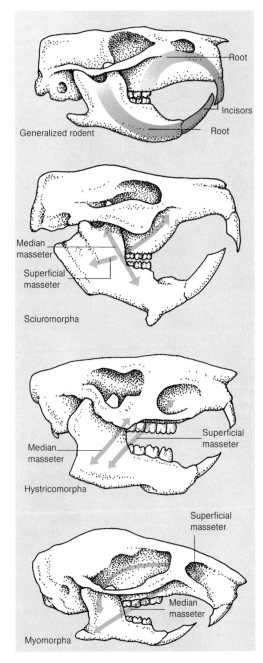

Generalized rodent

Root

Incisors

Root

Median masseter

Superficial masseter

Sciuromorpha

Superficial masseter

Median masseter

Hystricomorpha

Superficial masseter

Median masseter

Myomorpha

coat that is covered in coarse guard hairs. Beavers are excellent swimmers and they use their broad, flat tail as a rudder while they paddle with their webbed feet.

The springhaas, or Cape jumping hare, is the only species of the family Pedetidae. In most respects, it looks like a small version of the kangaroo, except that it has a bushy tail. It uses its long hind legs for jumping, but it also can move on all fours, like a rabbit. Like the jumping hare, the kangaroo rats in the family Heteromyidae are nocturnal burrowing animals, and their hind limbs are also modified for jumping.

The brightly colored scaly-tailed squirrels (Anomaluridae) and the North American flying squirrels are the only airborne rodents. They glide from tree to tree by means of a membrane that joins their limbs, called a patagium.

Mouselike rodents

The suborder Myomorpha contains more than 1,000 species, arranged in nine families. Their mouselike appearance and lack of premolars distinguish them from the sciuromorphs. The family Cricetidae contains about 570 species, including hamsters, lemmings, voles, and New World rats and mice. They live mainly above the ground, although some burrow and there are some semiaquatic species. Many have a thickset body and a short tail and legs.

The puna mouse is the only mammal to live at altitudes of about 16,400 feet (5,000 meters). It is found in the altiplana region of the Andes and is about 6 inches (15 centimeters) in length. The more familiar common hamster, which occurs across Europe into the Russian steppes, is a nocturnal species and may hibernate for a short period during winter. It stores food throughout the summer, some of which it carries to the winter burrow in its cheek pouches. This burrow is divided into separate compartments, for nesting, food storage, and body wastes. Each type of food, such as corn or potatoes, is stored in a different food compartment. Hamsters may store up to 200 pounds (90 kilograms) of food to see them through the winter. In summer, they dig another burrow, which they use for nesting. The golden hamster exists today only as a pet or laboratory animal.

Another member of this family, the true lemming, lives in the Northern Hemisphere, particularly in arctic regions. It is stocky, with very short ears and tail, and burrows in the soil during summer and under the snow in winter. When food supplies are plentiful, lemmings

The red tree squirrels of Europe, Asia, and the Americas inhabit large coniferous forests as well as mixed woodland and parks. The digits on their forelimbs are dexterous and used for holding food, at which they gnaw. Squirrels learn to crack open nuts only through trial and error. Eventually, by chiseling deeply with their incisors into the natural grooves of a nut's shell, they crack it easily.

breed rapidly. But every four to five years, their population increases to the point that there are too many individuals for the available food. The lemmings then start to migrate in search of new food sources. In Norway, the migrating lemmings keep moving until they reach the sea. This water barrier does not stop them and they plunge into the sea, where most of them drown. In the following three to four years, the numbers gradually rise to a similar level and the cycle repeats. Lemmings compete for the same food as reindeer and caribou and, at the peak of their population numbers, they deprive the large ungulates of food and cause many of them to starve. Some voles are also prone to similar cyclical fluctuations in numbers. The peak of population is known as a "vole year," during which there is a

Beavers are known for their building techniques. They not only build complex lodges (below) but also engineer dams and cut streams through woodland. The dams and lodges consist of a foundation of mud and stones on which the animal stacks brush and poles plastered with soggy vegetation and mud. The lodges are about 3 to 6 feet (91 to 180 centimeters) high with a diameter at the base of about 40 feet (12 meters).

Lodge   Ventilation shaft   Underwater entrance   Food store   Dam   Sleeping platform

Prairie dogs inhabit the North American plains in underground colonies called towns. These towns can contain about 1,000 animals and are divided into coteries, which have one male, three to four females, and several offspring. Sentries are posted at the entrance to the burrows, where they keep watch with a characteristic stance, standing upright on their hind limbs. If an intruder is spotted, they bark a warning. Prairie dogs are known to clear the surrounding land of plants they do not like in order to allow those they prefer to grow.

marked increase in the voles' predators, such as birds of prey.

One of the most unique rodents is the naked mole-rat (family Spalacidae) of eastern Africa. The naked mole-rat is the only known mammal whose social behavior resembles that of such insects as ants, termites, and bees. These rodents live underground in organized communities. Groups of similar individuals perform different functions in the colony, such as maintaining the tunnels and defending the colony from predators. One female takes the role of "queen" and produces young.

In addition to bamboo rats (Rhizomyidae) and jumping mice (Zapodidae), there are also the Old World rats and mice of the family Muridae. Many of them, especially the rats and the house mouse, have been introduced accidentally to all parts of the world, although they originated in the Old World. Most have long snouts and a long, naked, scaly tail. The black rat probably originated in southeastern Asia, but like the house mouse, it is now found throughout the world. Like other rats and mice, the black rat lives in close association with humans, inhabiting buildings and living off human food and garbage. These close as-

sociations often mean that the rats transmit diseases to humans.

The suborder also contains the dormice (Gliridae) and spiny dormice (Platacanthomyidae). They are mainly tree-dwelling and squirrellike, with a bushy tail. Like other leaf-eating rodents, they consume large amounts of berries, grains, and nuts in autumn to accumulate as much fat as they can. This energy reserve enables them to hibernate throughout the winter, during which they may lose half their body weight.

Porcupine-like rodents

Sixteen families comprise the suborder Hystricomorpha, and include porcupines, guinea pigs, and coypus. The porcupines make up two families: the Hystricidae, or Old World porcupines (found in Africa, Italy, and southern Asia), and the Erethizontidae, the New World porcupines of North and South America. In addition to hair on the body and tail, they have long, sharp, black-and-white quills on their back and flanks. The spines are loosely attached to the skin, and when the rodent attacks, it lunges sideways, then immedi-

The capybara, the largest of all rodents, can grow up to 4 feet (1.2 meters) long and may weigh over 100 pounds (45 kilograms). Its webbed toes make it a good swimmer. Capybaras graze near lakes and rivers and plunge into the water at any sign of danger.

ately withdraws, so that some of the quills or spines may become lodged in the victim.

The New World porcupines have shorter spines and some of them are barbed. They are tree-dwelling, and their feet are adapted for climbing. The sole is widened, and the first toe on the hind foot is replaced by a flexible pad. This group includes the cavies and guinea pigs (family Caviidae), the capybaras (Hydrochoeridae), and the chinchillas and vizcachas (Chinchilladae).

Lagomorphs

The members of the order Lagomorpha differ from rodents in that they have an additional pair of sharp incisor teeth in the upper jaw, which are situated behind the normal pair. In addition, these animals have a curious method of making the best use of plant food. To extract the maximum nutrient from their food, they first digest it in the stomach with the aid of bacteria and excrete it in the form of pellets. These pellets are eaten and pass through the digestive system once more—a process known as refection.

The family Ochotonidae consists of the pikas, or mousehares, found in North America and Asia. Of the Asian species, one lives on Mount Everest at an altitude of 5,740 feet (1,750 meters). Others live in deserts, on grasslands, in rocky regions, and in forests. Most of them inhabit areas with harsh winters. Pikas store food for winter. During the summer, they dry vegetation in the sun, gathering it into small piles that resemble haystacks.

The rabbits and hares make up the family Leporidae. They are native to most countries except Australia and New Zealand, although some have been introduced there by humans. Rabbits and hares are very similar, but they differ in several ways. Both groups are distinguished by their very long ears; short, upturned tail; and their hind legs, which are usually longer than their forelimbs. Most are active at night or at twilight and feed on bark, root crops, and grasses. The common names of rabbits and hares also sometimes make it difficult to distinguish them. The Belgian hare is really a type of rabbit. The snowshoe rabbit of the Arctic, which turns completely white in the winter, and the jack rabbit of desert areas

are, in fact, hares.

Hares are solitary except in the breeding season. They do not burrow, but make a depression in the grass, called a form, in which they rest during the day and give birth to their young, which are born with a set of teeth, a coat of hair, and their eyes open. Hares usually run swiftly to escape predators.

Rabbits escape predators by hiding. They build fur-lined nests, in which they give birth to naked, blind young. Unlike the solitary hare, rabbits live together in large groups of about 150 individuals. They inhabit lowlands and hills, generally where the soil is sandy.

The European wild rabbit has been introduced to most countries. It is extremely prolific—females are fertile at about six months of age and usually have five to seven litters a year, each with two to nine young. They are capable of becoming pregnant within 12 to 15 hours after giving birth. They are also voracious feeders, and these two facts make them serious agricultural pests when they are not controlled.

Old World porcupines, such as the Indian porcupine, have longer spines than their New World counterparts. This species has a short tail which, as well as being covered with sharp, barbed quills, has a cluster of hollow quills. These quills are shaken in warning and produce a sound similar to that of a rattlesnake. Porcupines orient their posterior toward predators in self-defense. Contrary to some assertions, porcupine quills are not shot or discharged at attackers. They are loosely attached and readily puncture predators that they contact.

Hares are notable mostly for their very long ears. Like rabbits, they eat bark, root crops, and grasses, but, in contrast to them, newborn hares are covered with hair, have developed eyesight, and teeth. They are also solitary animals, whereas rabbits are social.

Whales and dolphins

The cetaceans—whales, porpoises, and dolphins—have the most highly evolved brain of all aquatic animals and include some of the most intelligent animals on earth. Indeed, they are sometimes considered to be as intelligent as humans. Although cetaceans are descended from land-dwelling mammals, they have nevertheless become so well adapted for an aquatic life that they cannot leave the water. Except for the few species of dolphins of the family Platanistidae, which live in rivers and lakes, all cetaceans are marine mammals.

The order Cetacea divides into two main suborders, according to method of feeding. The suborder Odontoceti, which has seven families, is made up of the toothed whales. The suborder Mysticeti, the baleen whales, has three families.

The common names of these animals are confusing. Most large species are called whales and most of the smaller ones are called dolphins. The killer whale, however, is a member of the dolphin family (Delphinidae), yet there are several species of whales considerably smaller than it.

A further inconsistency is that in the United States, most dolphins are called porpoises, whereas in Europe, this name is reserved for six species of small, blunt-faced dolphins of the family Phocaenidae.

Anatomy and physiology

Of all the mammals, whales and dolphins are the most highly adapted for life in water. The outline of their body has become streamlined, and their forelimbs have become broad flippers, which they use for steering and balance. Although these flippers still have all the same bones of the vertebrate forelimb, they have been modified. The hind limbs have disappeared, but cetaceans still have a reduced pelvis that in the male supports part of the reproductive apparatus. The dorsal fin and the tail are folds of skin that are not supported by any skeletal elements. Cetaceans' smooth outline is partly due to the absence of fur—cetaceans have only a sparse covering of hair.

Cetaceans propel themselves through the water by up-and-down movements of the broad horizontal tail flukes. The body is protected by layers of fat, or blubber, which act as an efficient insulator to keep the body warm and also serve as an energy store. The presence of large amounts of fat in the surface tissues allows subtle variations in the animals' shape, which reduce drag and make deep diving easier.

Cetaceans usually show only a small part of their back when they come up to breathe, although even the largest ones can leap clear of the water, a behavior known as breaching. The nostrils are on the top of the head, forming a single blowhole in toothed whales, and two blowholes in baleen whales. As a cetacean surfaces, it blows, sending up a distinctive spout of watery spray. This jet is a cloud of droplets caused by water vapor in the breath condensing as the pressure suddenly drops when the animal shoots up to the surface.

The circulatory and respiratory systems of cetaceans are well suited for deep diving. Cetaceans carry a relatively small proportion of their oxygen intake in their lungs. Evidence suggests that the lungs are emptied before a deep dive and that oxygen is stored elsewhere, most of it combined with hemoglobin and myoglobin in the blood and tissues. Because there is not enough oxygen to supply all body functions during a long dive, cetaceans compensate by directing oxygen-rich blood toward the brain and nerves, where it is most needed.

These animals also have evolved some muscles that can work for short periods anaerobically (without oxygen), although they must be replenished with oxygen soon afterward. The reduced amount of oxygen cetaceans take on deep dives, and the diminished rate of blood flow to the tissues prevents the problem of nitrogen coming out of solution as bubbles in the blood when these animals return to the

The bottle-nosed dolphin occurs in all the world's oceans, but is most common along the Atlantic coast of the United States. It eats fish and, like other true dolphins, can often be seen swimming alongside or in front of ships.

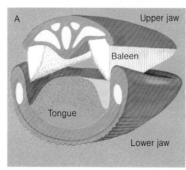

A baleen whale filters the tiny organisms in plankton from seawater using baleen plates that hang down from the upper jaws *(right, A)*. The blue whale *(far right, B)*, is a baleen whale. It is the largest mammal that has ever lived and is now in danger of extinction.

A   Upper jaw

Baleen

Tongue

Lower jaw

B

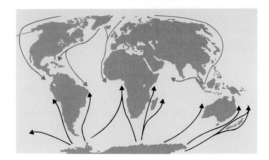

A young humpback whale swims above its mother. It is fed on mother's milk for five to ten months after birth, but does not grow to full size until it is 10 years old. These whales migrate annually from Arctic and Antarctic waters *(map, left)* to spend the winter in warm tropical seas.

surface—a dangerous condition for human divers called the bends.

Cetaceans have a highly developed sense of touch. They have very little sense of smell, and a variable ability to taste. Whales and dolphins, in fact, have no sense of smell. Many species have good eyesight, but vision is not very useful because little light penetrates the depths of the sea.

Because sound travels well in water, cetaceans depend chiefly on their sense of hearing. Whales and dolphins are able to produce a range of different noises, with which they communicate and find their way. Schools of dolphins chatter continually to one another, using clicking and whistling sounds, and the "songs" of the humpback whale are known to carry hundreds of miles. Such calls inform the cetaceans of the identity and mood of their fellows. Distress calls, for instance, seem to summon other cetaceans to an injured individual. Some species of toothed whales, especially dolphins, are known to use echolocation. Like bats, these creatures emit streams of high-frequency clicks and ultrasonic squeaks and use the returning echoes to locate prey and to orient themselves. There is evidence that dolphins and killer whales have a language, and an ability to reason, learn, remember, and use information.

Teeth and feeding

The whales of the suborder Mysticeti have no teeth. Instead, they have a series of comblike baleen plates made of keratin (the same substance as fingernails) hanging down from the upper jaw, or palate. Their food consists of plankton, the mass of tiny organisms that drifts at or near the surface of oceans, and also sometimes small fish. The whale draws water containing the food into its mouth and then squirts it out through the baleen plates, which act as a strainer, keeping the food in the mouth.

Toothed whales include all nonbaleen whales. They have numerous conelike teeth. Some toothed whales, such as the sperm whales (family Physeteridae), have teeth in their lower jaw only, and the beaked and bottle-nosed whales (family Ziphiidae) usually have only one or two pairs of teeth showing, but even these may never break through the gums. The male narwhal has a single tooth that has become a long, twisted tusk on one side of its head.

The teeth of most toothed whales are adapted for holding prey but not for chewing. As a result, they swallow their food whole. Most toothed whales feed on fishes, squid, and cuttlefish, but certain groups of killer whales specialize in hunting seals.

Reproduction

Cetaceans breed seasonally, mating during spring and early summer. They produce single offspring, and birth takes place from 10 to 12 months after conception in most species, although sperm whales have a very long gestation period of some 16 months. One or more females may help a mother cetacean give birth. Calves are born under water and may be as much as a third of the mother's size at birth. A calf emerges tail first and is pushed to the surface by its mother to take its first breath and to suckle from the teats, which lie in the folds of blubber on each side of the reproductive opening. The milk is 50 per cent fat and is very rich in protein and minerals.

Baleen whales seem to breed only every two years. As a result, the numbers already severely depleted by the commercial whaling industry will take a long time to recover. In addition, large whales start to reproduce only when they are about 12 years old, which obviously affects the replacement of individuals in nature.

Carnivores

The carnivores (order Carnivora) are the chief flesh-eating mammals, although hunting as a way of life is shared by others, such as seals and some whales. But the name *Carnivora* is misleading because at least two species—the red, or lesser, panda and the giant panda—eat mostly leaves, and some bears eat very little meat. Also, many of the smaller carnivore species are actually omnivorous or insectivorous. The order, which contains seven families and more than 100 genera, is distributed throughout the world, except for Antarctica and some islands.

Most carnivores live on the ground or in trees, but there are also aquatic species, such as the sea otter. Although social patterns vary widely among species, most are fiercely territorial. The two most prominent groups, the cats and their relations and the dogs and related species, have developed different hunting techniques. Dogs tend to run down their prey and many are social and hunt in packs. With the notable exception of African lions, cats tend to be solitary ambushers.

Carnivore anatomy

The long, flexible body of the hunting animals and their way of life are adapted to their predatory existence. Terrestrial carnivores are surefooted and agile, and can run swiftly. Most species naturally stand on the tips of their toes with four or five toes touching the ground, and many have claws. The claws help grip the ground and grasp prey. They are also useful for digging and scratching. Some species, such as dogs, cats, and hyenas, have a dewclaw on each forefoot, which represents a toe that no longer touches the ground.

One of the distinguishing features of carnivores is their long, pointed, canine teeth, which they use for stabbing and holding prey. The skull of a carnivore is strong, and powerful muscles work the jaws. In some species, there is a ridge of bone on top of the cranium called a sagittal crest, which serves to increase the anchorage for the upper end of the jaw muscles. The lower jaw can only open and shut, with little ability for sideways grinding movements of the teeth.

In proportion to their body size, carnivores have a large, well developed brain that makes possible the intelligent behavior needed for hunting. Their senses are efficient, enabling the animals first to locate prey at a distance and then to guide the attack. Carnivores have forward-facing eyes that help in judging distance accurately when they spring on prey. The eyes of many carnivores have an internal reflector, called the tapetum, which increases the eyes' sensitivity at night. A structure called the lucidum causes carnivores' eyes to shine when a light is directed at them.

Carnivores have a fine sense of smell, important both for trailing prey and for communication. Scent released from the anal and other glands conveys information about an animal's identity, sex, and territorial ownership.

Most carnivores breed only once a year although a few species, such as some weasels, reproduce twice a year. Litters vary in size from the single offspring of bears to the 12 or more of skunks. The young are usually blind at birth and are necessarily dependent on parental care for some time. In captivity, some carnivores have been known to live for more than 30 years.

The dogs

The dogs (family Canidae) are generalized hunters. The family includes wolves, coyotes, jackals, foxes, and the Cape hunting dog. They are the most vocal of the carnivores, capable of a variety of barks, howls, and whines. Those members of the family that hunt in packs can bring down large animals, but the solitary hunters usually live on small rodents, insects, or birds.

The wolf looks like a heavily built German shepherd dog. Until persecution by humans drove it from much of its former range, it

Carnivores have four kinds of teeth: incisors, canines, premolars, and molars. They use incisors for biting off flesh and canines for stabbing and grasping prey. The premolars and molars, for slicing and chewing, show considerable modification according to diet. Bears chew their food thoroughly and have flat cheek teeth to crush vegetable matter. Cats (except for lions) and dogs have pointed cheek teeth called carnassials, which slice and tear meat into chunks.

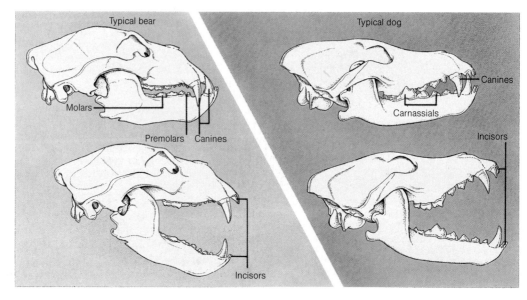

could be found over most of the Northern Hemisphere. The basic unit of wolf society is a female with her offspring. When the cubs mature, they stay with the mother and hunt with her. The male stays with his family and helps feed the cubs. Lone wolves eat anything they can obtain, but most wolves form packs of 8 to 20 animals, which are highly efficient hunting groups capable of bringing down prey as large as an elk.

The coyote resembles a small wolf. It is found throughout North America and lives in pairs or family groups that feed on anything from deer to trash from humans.

Several species of jackals live throughout Africa and southern Asia. Jackals usually live in pairs that cooperate in hunting, and young from a previous litter may also help to feed pups. Jackals sometimes gather in packs to prey on large animals or to scavenge.

The red fox has as extensive a range as the wolf, and has been introduced into Australia as a quarry for hunters. Unlike the wolf, the fox has survived continuous persecution and has even taken to living on the fringes of towns and cities, where it scavenges on people's garbage. Some foxes are wanderers without a permanent home, but most live in pairs in territories, although they hunt separately.

The Arctic fox has two forms of seasonal color change. It has either white fur in winter, which turns brown in summer, or a blue-gray winter coat that becomes darker in summer. This variation seems to reflect the harshness of the climate and the environment.

The hot, dry parts of Africa are the home of the bat-eared fox and the fennec. Both species have large ears that act as radiators to help keep the animal cool and as sound receivers to increase the sensitivity of hearing.

The Cape hunting dog of the African savanna is exceptionally social. It lives in packs of up to 20 that cooperate to hunt zebras and antelopes. The members of the pack share the food and, when a female has pups, the other dogs bring food back to the den.

The bears

The bears (family Ursidae) are the largest mem-

bers of the carnivore order and are found in the Northern Hemisphere and in parts of South America. Zoologists generally recognize seven species of bears—big brown bears, American black bears, Asiatic black bears, polar bears, sun bears, sloth bears, and spectacled bears. For many years, zoologists hotly debated whether the giant panda should be classified in the bear family or the raccoon family. However, recent comparisons of panda DNA to that of other bears confirms that pandas are more closely related to bears than to raccoons.

Bears are heavy-bodied, and most species live in forests. They are usually omnivorous and have blunt teeth. They eat a variety of plant foods, such as fruit, grass, nuts, and acorns, but eagerly feed on fish and other small animals when they are available. They are especially fond of ants, termites, and

The foxlike fennec lives in the African desert, feeding on a varied diet of rodents, reptiles, and insects. It also eats fruit, such as dates. The fennec rarely drinks, but relies instead on the body fluids of its prey to provide it with moisture.

Cape hunting dogs live in the open on the grasslands of southern Africa, forming packs to hunt herbivores, such as wildebeest *(below)* and antelopes.

An Alaskan kodiak bear, a type of brown bear, prepares to feed on a salmon. Although bears are grouped with the carnivores, much of their diet consists of fruit and grasses, and so they actually are correctly regarded as omnivorous.

grubs. Bears are solitary animals, although several may gather around a good source of food such as a garbage dump or a shallow salmon run.

Bear species that live in cold places spend most of the winter in a state similar to sleeping. Zoologists disagree about whether this winter sleep is actually hibernation. Some point out that a bear's body temperature, unlike that of other hibernating animals, does not drop significantly during winter sleep. In addition, a bear wakes up easily, and may become fairly active on mild winter days. As a result, many zoologists prefer to refer to a bear's sleep period as winter lethargy or incomplete hibernation.

Most bear cubs are born during the winter sleep period of the mother bear. Twins are most common, but the number of young born may vary from one to four. The cubs are extremely small at birth, weighing less than 1/350 of their mother's weight.

Kinds of bears

The big brown bear weighs more than 1,500 pounds (680 kilograms). It lives in the northern forested areas of North America, Europe, and Asia. The grizzly bear is a large subspecies of the brown bear, but is less common. Another North American bear is the black bear, which

is smaller than the brown bear. It is an agile climber even when adult. The spectacled bear is the only true bear in the Southern Hemisphere. The Asiatic black bear is also called a moon bear because of a white, crescent-shaped mark on its chest. The sloth bear of India and Sri Lanka has a long snout with mobile lips that it uses to suck termites from their nests. The female carries her young on her back when they first leave the den. The smallest bear is the sun bear, also called a Malayan bear, of Burma, Malaysia, and Borneo. It is an excellent climber and has a short coat that is less shaggy than that of other bears.

The polar bear differs from other bears in that its diet consists almost exclusively of meat. This bear lives in the Arctic, where other types of food are unavailable. The polar bear spends most of its time on pack ice, where it preys mainly on seals. It is a strong swimmer but never hunts in water. The bear either pulls seals and fish out of the water or catches seals on the ice. Polar bears also occasionally eat birds, hares, and berries. The thick white fur of the polar bear covers even the soles of its feet, giving added grip on the ice as well as insulation from the cold.

Giant pandas live in the remote bamboo and rhododendron forests of southwestern China and eastern Tibet, where they feed almost exclusively on bamboo shoots. They have a woolly white coat with black legs and shoulders, black ears, and black patches around the eyes. Adults measure up to 5 feet (1.5 meters) long and weigh 150 to 350 pounds (70 to 160 kilograms). A unique feature of pandas is a sesamoid bone that rises from the wrist and acts as a sixth finger on the forepaw, used to hold the shoots. Pandas have broad teeth for chewing tough shoots into a pulp, and a muscular stomach for aiding digestion.

The raccoon family

The raccoon family (Procyonidae) includes the raccoons, cacomistles, kinkajous, ringtails, and red, or lesser, pandas. Most are small animals that spend much of their time in trees. Except for the red panda, all raccoons live in temperate and tropical areas of the Americas and eastern Asia. Some procyonids have long, ringed tails. That of the tree-dwelling kinkajou is prehensile. The diet of raccoons includes a variety of small animals, although the kinkajou and olingo eat mainly fruit. Most procyonids are nocturnal, except for coatis, which usually move about during the day. Most raccoons also live in pairs or groups. The coatis are the most social of this group, living in large bands.

The raccoon is found throughout North America and in parts of Central America. Like the fox, it has adapted to living in populated areas and scavenges in trash cans, but in wilder regions it frequently feeds along the water's edge on shellfish, crayfish, and frogs.

Like the giant panda, the red panda also lives in bamboo forests, from China to Nepal. It has a similar sixth finger for handling bamboo but frequently also eats acorns, roots, fruit, and small birds.

The weasel family

The family Mustelidae, with 67 species, is the

The hind limbs of a dog *(below)* and a bear *(below, right)* reflect their ways of walking. Dogs are digitigrade—that is, they walk on their fleshy toe pads. Bears are plantigrade. They walk on the sole of the foot with the heel in contact with the ground. A dog has blunt claws on its four toes, whereas a bear has sharp curved claws on each of its five toes.

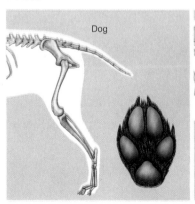

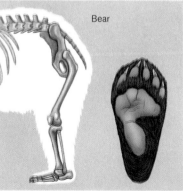

Dog

Bear

largest and most varied of the carnivores. Most mustelids have long, slender bodies and short legs and are hunters of earthworms, frogs, lizards, snakes, and warm-blooded prey. From this basic pattern they have developed a variety of life styles. The martens are tree-climbers, the minks and otters are aquatic, and the badgers are stocky digging animals. Weasels' diets also vary among species. For example, the tayra of Central America and South America eats fruit, small animals, and birds. The larger wolverine, or glutton, is a powerful scavenger of the high northern latitudes that also can kill deer and cattle weakened by harsh weather.

Weasels are lithe, fast-moving animals and are small enough to follow mice and voles down their burrows. In the northern parts of the weasel range—North America, Europe, Asia, and Indonesia—stoats and weasels grow a white coat in winter, which makes them inconspicuous in snow. Minks and polecats are larger members of the family. And the American least weasel is the smallest flesh-eating animal in the world.

The black-footed ferret is one of the rarest carnivores. It lives on the North American prairies and preys almost exclusively on prairie dogs. It was virtually extinct in the wild because farmers and ranchers had exterminated prairie dogs in many areas. However, black-footed ferrets are making a comeback, thanks to reintroduction of zoo-born animals into the wild.

Most carnivores have anal glands that secrete a fluid used for marking territory. But the best-known example is the skunk, also a mustelid, which uses the fluid for defense. The glands also have been modified for the same purpose in the Asian stink badger, the African striped weasel, and the African zorilla.

The badger of Europe and Asia is one of the few social mustelids—and one of the largest. It lives in family groups called clans. The clan territory centers on the animals' burrows, or setts, which may form an extensive underground network. Badgers are nocturnal and feed mainly on earthworms and insects. The American badger, however, is more solitary and its diet is mainly rodents. The honey badger, or ratel, is known for its association with the honey guide, a bird that leads honey-eating mammals to beehives. The ratel opens the nest to devour the contents, and the honey guide feeds on the scraps.

Some of the largest mustelids are the otters. Eighteen species live in fresh water, but some

A **striped skunk** raises its tail as a warning (A) when threatened. Before spraying, it gives warning by stamping its front feet or hissing and growling. It arches its back and may raise itself on its front paws (B). If the threat persists, the skunk turns and ejects two jets of foul smelling fluid from its anal glands (C).

The raccoon is an opportunistic feeder. Those that live in remote areas eat frogs and fish, while those that live near populated areas scavenge in trash cans and garbage.

The stoat, or ermine, as it is called when in its white winter coat, can run extremely fast, chasing down its prey of small rodents, rabbits, and squirrels.

Eggs form part of the diet of the banded mongoose of Africa together with insects, snails, and mice. It also eats small reptiles and fruit. Mongooses roam in bands, taking temporary shelter in the abandoned nests or burrows of other animals.

tails. Some, such as the genets, are agile, cat-like tree climbers, whereas the otter civet and crab-eating mongoose hunt in water.

Viverrids prey mostly on small animals, though some also eat fruit. Mongooses are known to attack snakes, using their speed and agility to confuse the snake and avoid its counterattacks. They have been introduced to several parts of the world to destroy poisonous snakes and rats, but cause destruction among native wildlife and also raid poultry pens, mainly for eggs. Most viverrids are solitary and active by night, because although they are good fighters, they are small enough to be attacked by such birds of prey as eagles, which hunt by day. Some species of viverrids that are active during the day live in troops and cooperate to watch for danger and drive away large predators.

The hyenas

Hyenas (family Hyaenidae) have had a bad reputation as cowardly scavengers, but are now known to be fierce predators as well as eaters of carrion. The three species that live in Africa and southwestern Asia are strong runners and hunt in packs, chasing herds of antelopes or zebras until a victim can be caught and pulled down. Their powerful jaws and large, sharp teeth can crunch even the largest bones. Hyenas have a big head and well developed forelegs. They have a characteristic trot, but are also able to run at high speeds. Spotted hyenas are as noisy as many dog species, making howling and "laughing" sounds.

The African aardwolf is a type of hyena, but it has weak jaws, small teeth, and lives mostly on termites and other insects.

The cats

The family Felidae consists of six genera and 36 species. The main genus, *Felis*, with 25 species, includes the mountain lion, or puma; the ocelot; the serval; and many species of smaller wild cats in addition to the domestic cat. The other main genera are *Lynx*, with four species, including the Caracas and the bobcat,

river otters and the sea otter frequent the sea and shoreline. Otters swim with powerful undulations of their bodies and broad tails and steer with their webbed feet. They can dive for several minutes in search of fish and other aquatic animals, closing their ears and nostrils while submerged. The sea otter, which lives off the west coast of North America, feeds on crabs, sea urchins, and shellfish. It smashes them open by banging them on a stone, which it carries on its chest while floating on its back. It also carries and nurses its young in this position. Unlike most marine animals, sea otters do not have a heavy insulating layer of fat. Instead, they rely on the protection of a layer of air trapped by their long, soft fur.

The mongoose family

The mongooses and civets (family Viverridae) resemble weasels and, indeed, occupy similar ecological niches in Old World tropical regions. Many have spotted coats and ringed

Tiger *Panthera tigris*

Lion *Panthera leo*

Leopard *Panthera pardus*

Young cheetahs feed on an antelope that has been run down and killed by their mother. The fastest of all mammals, cheetahs can run in short bursts of up to 70 miles (110 kilometers) per hour for distances up to several hundred yards or meters. They often chase their prey for 450 yards (411 meters) or more before springing on it.

and *Panthera,* which also has four species—the jaguar, the leopard, the tiger, and the lion.

Compared with other carnivores, the cats have a short muzzle and a broad, rounded head. With the exception of the lion, the fur is soft and often marked with spots or stripes. Cats are specialized hunters with lithe, compact bodies and large, sharp, scissorlike cheek teeth, called carnassials. The whiskers are long and sensitive, and cats have acute sight, hearing, and sense of smell. Cats vary widely in size and appearance. The smallest wild cats are about the same size as the domestic cat, whereas the largest species, the tiger, has a body length of more than 9 feet (2.8 meters), or 10.5 feet (3.2 meters) including the tail.

Most cats are nocturnal, and all but the lion hunt alone. Lions live in prides of several females and their cubs, with one or more males. Several members of the pride usually hunt together. They lie in ambush or sneak forward slowly toward their prey until they are close enough to leap up and catch it, bringing it down with their claws and a bite that breaks the victim's neck. Unlike most carnivores, all cats except the cheetah have retractile claws. They extend these claws to help grasp and slash prey, and withdraw them into sheaths when they are not in use. The cheetah's blunt, nonretractile claws give it a good grip on the ground as it sprints at speeds up to about 70 miles (110 kilometers) per hour. Its light build and flexibility help it to turn fast, and its long tail provides steering and balance.

The big cats—the lion, tiger, leopard, and jaguar—are also characterized by their roar, which is heard at night. The roar is made possible by an additional ligament in the throat that is attached to the hyoid bone.

Cheetah *Acinonyx jubatus*

Lynx *Lynx canadensis*

Fishing cat
Felis viverrina

Puma
Felis concolor

European wild cat
Felis sylvestris

Enormous variety in size and build of the cats becomes evident when species are compared. The head and body of a lion is about 8 feet (2.4 meters) long, and a tiger is even larger at up to 9 feet (2.8 meters) long. The leopard, cheetah, and mountain lion are similar in size—5 feet (1.5 meters)—whereas the lynx and European wildcat are slightly smaller. The body length of the fishing cat is about 28 inches (70 centimeters). With the exception of the cheetah, these cats stalk their prey or spring from ambush.

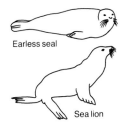

Earless seal

Sea lion

An earless seal cannot turn its hind flippers forward and so is much less mobile on land than the sea lion.

Elephant seals are the largest of the seal group. The males weigh up to 8,000 pounds (3,600 kilograms) and can measure up to 21 feet (6.4 meters) in length. Their name is derived from the bladder on their nose, which can be 15 inches (38 centimeters) long. These animals are polygamous; the bulls become belligerent in the breeding season and challenge each other's dominance to acquire a harem.

Seals

The seals are marine mammals. They have been classified as a suborder in the order carnivora, but some zoologists consider the seals a separate order, Pinnipedia. The name comes from Latin words meaning "fin-footed."

The order is divided into three families—the earless seals of the family Phocidae; the sea lions and fur seals, also known as the eared seals, of the family Otariidae; and the walrus, of the family Odobenidae. Seals are found along most coasts but some of the biggest concentrations are in Arctic and Antarctic waters.

General features

Pinnipeds spend most of their lives in water, although they go on land to bear and rear their pups and to molt. Like the cetaceans, they are well adapted for an aquatic life. Their bodies are streamlined and torpedo-shaped. They are padded with blubber that acts as an energy store and that provides insulation. Their limbs have been modified into flippers for swimming. Pinnipeds have either tiny external ears or none at all, and slitlike nostrils. The ears and nostrils are closed when the animals are submerged, but the eyes remain open and are efficient under water. Whereas walruses are almost bare-skinned, most pinnipeds are covered with short hair.

All pinnipeds are well adapted for diving.

When they plunge, their heart rate immediately drops from 50 to 100 beats per minute to 10 or less. The remaining blood flow is directed mainly to the brain. Like whales, pinnipeds can survive without breathing for much longer periods than land mammals. The record for diving is probably held by the Weddell seal, recorded at a depth of some 2,360 feet (719 meters) for up to 43 minutes.

Male pinnipeds are much larger than the females. A large bull southern elephant seal weighs up to 8,000 pounds (3,600 kilograms), four times more than the cow. The life span varies according to species, from about 25 years to 40 years or more.

Most pinnipeds feed on fish, but eared seals also eat crustaceans and marine mollusks, such as squid and octopus. Other pinnipeds, such as the leopard seals also eat the pups of other seals, and a genus of sea lion occasionally eats penguins. The diet of the walrus consists chiefly of clams and oysters.

Cows give birth to one pup a year, usually at a traditional breeding space called a rookery, where large colonies of seals gather. Most rookeries are on islands, but rookeries of northern fur seals are large beach areas. More than 150,000 seals may gather at a single rookery. The bull seals at these rookeries are usually polygamous—that is, one bull mates with several cows—and arrive there before the cows do to establish territories, fighting off other males.

After a gestation period of about 12 months, the pup is born. Earless seals suckle their pups for a short time. The harp seal, for example, suckles its young for only nine days after birth. The pups develop very rapidly and during this period the cow does not feed, drawing on her reserves of blubber to produce milk. The pups of some species are able to swim within hours of birth, whereas others take several weeks. Eared seals rear their pups very slowly. Galapagos fur seals, for example, are not weaned until they are two or three years old.

A few days after a cow seal has given birth, she mates again, but the implantation is delayed for a few months. This allows the female to recover from the strain of feeding her pup. It also means that pups are always born at about the same time of year, which is an important consideration for migratory species.

The hind limbs of earless seals are contained within the body. The small forelimbs are placed well forward. Earless seals move with difficulty on land, dragging themselves onto their front flippers (A) and wriggling forward (B). They then launch themselves forward from their hind flippers (C) and collapse into a new position (D).

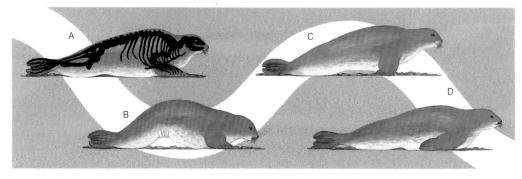

A

B

C

D

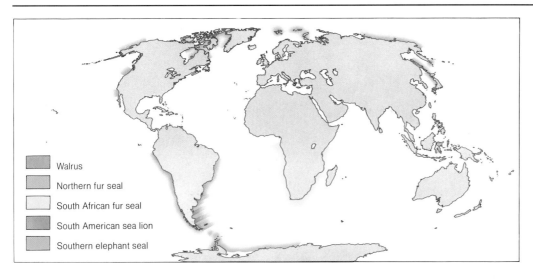

Some seals migrate. Walruses live in open Arctic waters, migrating south in winter before the ice closes in. Northern fur seals gather to breed at islands in the North Pacific, Okhotsk Sea, and Sea of Japan, and then disperse south. South African fur seals, South American sea lions, and southern elephant seals are not really migratory, though the latter have been sighted as far north as the island of Saint Helena, which lies about 1,200 miles (1,930 kilometers) southwest of Africa.

- Walrus
- Northern fur seal
- South African fur seal
- South American sea lion
- Southern elephant seal

The earless seals

The 18 species of earless seals live mostly in the Arctic and Antarctic, but some inhabit temperate waters. The main exceptions are the monk seals, which live in warm seas, and some seals that inhabit freshwater lakes, such as Lake Baikal in Russia and the Caspian Sea.

Pups of most species are born with a soft, dense, woollike white coat, which is replaced by the stiffer adult coat before they start to swim. The white coat serves as camouflage on the ice and its thickness provides good insulation. Common seal pups, however, are born with their adult coat.

The crabeater seal is the most abundant of all earless seal species, with a population of 15 million. Despite its name, it feeds on tiny crustaceans known as krill, which it strains from the water through its teeth. Young crabeaters are hunted by leopard seals and many bear large scars from these attacks.

They are clumsy on land, where they move by pushing the ground with their front flippers. In water, however, they are very graceful, swimming with side-to-side movements of the hind flippers and using the foreflippers for maneuvering at low speed.

The eared seals

Of the 16 species of eared seals, five are sea lions and eleven are fur seals. They are found in polar, temperate, and subtropical areas. Eared seals get their name because of their tiny cartilaginous earflaps, which earless seals do not have. Eared seals are long and slender. Unlike earless seals, which look like they have no tail at all, the eared seals have short, but developed tails.

Fur seals have a thick undercoat of soft fur, protected by longer, coarser hair, whereas sea lions have only one layer of coarse hair. Sea lions are the biggest eared seals, have a heavier muzzle and head, and also a greater disparity in size between the sexes than other seals.

The eared seals and the walrus swim at high speeds by undulating their bodies, with their foreflippers tucked well back. The northern fur seal can swim 10 miles (16 kilometers) per hour. Eared seals are more agile on land than are earless seals because they can turn their hind flippers forward. In this way, they can lift their body off the ground and move forward.

The walrus

The single species of walrus lives in the Arctic Ocean. It migrates south in winter, riding on the ice floes, and returns in the spring. It has a thick body and a tough, hairless skin that covers a layer of blubber 1 to 6 inches (2.5 to 15 centimeters) thick. Both sexes have a pair of tusks, which are long upper canines. They use these tusks to stir up food from the seabed, to fight with other walruses, to defend themselves against polar bears, and with the help of the fins, to clamber up onto the ice. Walruses feed on such shellfish as clams and oysters. The bulls can measure 10 to 13 feet (3 to 4 meters) long and weigh as much as 3,000 pounds (1,360 kilograms). Despite their size, they are excellent swimmers and can reach speeds of 15 miles (24 kilometers) per hour.

Adult walruses have long powerful tusks that are actually canine teeth. The males brandish these tusks in threatening displays when competing with each other for females.

Indian elephant

Dugong

Rock cony

Aardvark

The subungulates, shown here to scale, have a wide variety of forms and sizes. But they are classified together mainly because of their teeth.

Aardvarks and subungulates

Plant-eating mammals that possess hoofs are referred to as ungulates. But many nonhoofed animals have evolved from primitive ungulate ancestors and have features that are similar to the ungulates. Consequently, they are called subungulates. Subungulates include the aardvark, elephants, manatees, and conies.

Aardvarks

As the only surviving species of primitive ungulates, the aardvark is classified in an order by itself—the Tubulidentata. The name relates to the aardvark's unusual molar teeth, which contain many tubular pulp cavities.

Aardvarks are found in the savanna regions of Africa. Their name means "earth pig" in Afrikaans, a Dutch dialect spoken in South Africa, and refers to the aardvark's appearance and burrowing habits. These animals are stocky and powerful, weighing up to 140 pounds (64 kilograms). They have large ears that flatten to keep out the soil when they are burrowing. They have strong limbs and a thick, muscular tail. Aardvarks resemble the anteaters of South America in several ways— both have an elongated snout, no teeth, and a long tongue that they can extend from their mouths. These features reflect parallel adaptations to a similar diet.

Aardvarks feed mainly on ants and termites, using their strong claws to break open sun-baked ant and termite hills. They also collect the insects as they swarm across the ground. An aardvark's tongue is covered with sticky saliva, to which the insects adhere, and the aardvark can extend its tongue up to 18 inches (46 centimeters). Their thick skin appears to protect them from the bites of termites.

They are nocturnal, traveling up to 19 miles (30 kilometers) every night in search of termites. During the day, they sleep in their burrows. They usually have single offspring, but occasionally produce twins.

Elephants

The present order of elephants (Proboscidea)

is a mere remnant of its former size. In the Ice Age, numerous species, which included the mammoths, were spread throughout the world. Today, only one family remains (Elephantidae), containing two species—the African elephant and the Asiatic, or Indian, elephant. They are easily distinguished from each other because the African species has larger ears than its Asian counterpart and grows bigger. The African elephant is the largest living land animal, weighing up to 6 short tons (5.4 metric tons). To support their weight, the limbs of elephants have become pillarlike, with each bone resting directly on the one below. Each foot has a thick pad of tissue that acts as a cushion against the elephant's weight.

Elephants often consume more than 770 pounds (349 kilograms) of vegetable matter per day and may spend up to 16 hours a day feeding. Elephants use their trunk, which is an extension of the nose and upper lip, to collect food and water. At the trunk tip are fingerlike projections (one in the Asiatic elephant and two in the African), which allow the trunk to pick up objects as small as peanuts. Most of their food is woody and fibrous, and is broken down by very large molar teeth with jagged ridges. Each jaw has six molar teeth per side, but normally only one tooth is present at a time. As the tooth wears away, another one erupts from the back of the jaw. Once the elephant loses the last tooth, it has difficulty feeding and may starve. The much sought after ivory tusks of elephants are not canines, as is often thought, but rather well developed upper incisors. Tusks of more than 10 feet (3 meters) long are known to have been taken from African bull elephants.

Conies

Conies, or hyraxes, (order Hyracoidea) resemble large, gray-brown guinea pigs. But they are more closely related to elephants than to rodents, particularly in the arrangement of their teeth. Conies have well developed grinding cheek teeth, and upper incisors that grow

Conies, or hyraxes, are agile climbers, both on rocks and in trees. This rock hyrax keeps a lookout for possible predators while the rest of its colony basks in the sun.

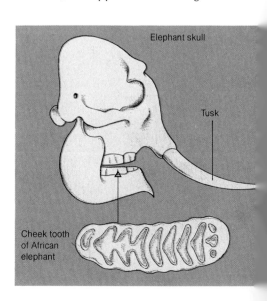

Elephant skull

Tusk

Cheek tooth of African elephant

130

continually and protrude from the mouth. These teeth are equivalent to tusks in elephants. The cony digestive system is unusual in that it has two appendixes.

Rock conies are highly sociable animals and they live in large colonies. They are active during the day and often bask in the sun, while older members of the colony act as lookouts. These sentries give a shrill alarm call if a predatory animal, such as a leopard, bird of prey, or python, approaches. Tree conies are less sociable than rock conies and are nocturnal, spending the day in tree holes or foliage.

All conies are highly vocal and use a wide range of croaks, cries, and alarm calls. All species also have a gland on the back, which is marked by a patch of different colored hair. This gland is exposed whenever the surrounding hairs stand erect—during the mating season or when the cony is frightened, for example. The gland is used to mark territory.

Manatees, dugongs, and sea cows

Despite their large, blubber-filled bodies, this group (order Sirenia) is thought to have been the source of the mermaid stories. These aquatic mammals live in coastal waters and large river systems in tropical and subtropical areas. Weighing up to 1,500 pounds (680 kilograms) and ranging in size from about 6 to 13 feet (2 to 4 meters) long, they never venture ashore. Today there is one genus of manatees (*Trichechus*), which contains three species, and one genus of dugongs and sea cows (*Dugong*), of which only dugongs exist. Sea cows became extinct in the mid-1700's after sailors killed them off for food.

Sirenians feed on seaweeds or freshwater plants. In Guyana, manatees have been used to control water weeds, which would otherwise choke up the rivers. The dugong has tusklike incisors and three cheek teeth on each side, whereas the manatee has no incisors, but up to 10 cheek teeth on each side, which are replaced and move forward continually.

The forelimbs of manatees form paddles, as in many other aquatic mammals. There are no hind limbs, and the tail has a horizontal fluke similar to a whale's. Manatees can remain submerged for up to 15 minutes at a time.

African elephants roam in herds consisting mainly of females and juveniles. After a gestation period of 18 to 23 months, pregnant females leave the herd to give birth. Like other mammals, the newborn young are fed on milk. The female has two nipples, which are located between her front legs.

The forelimbs of manatees form paddles, as in many other aquatic mammals. There are no hindlimbs, and the tail has a horizontal fluke like a whale's (but unlike seals and sea lions). Manatees can remain submerged for up to 15 minutes before having to surface to breathe.

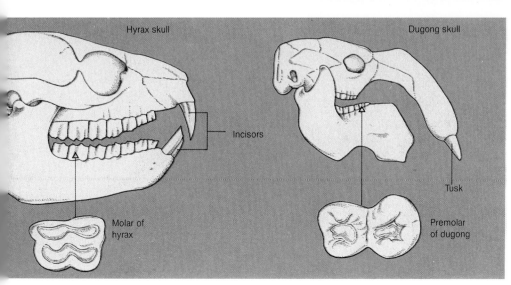

Hyrax skull

Incisors

Molar of hyrax

Dugong skull

Tusk

Premolar of dugong

Elephants, conies, and dugongs all eat plants and have a series of cheek teeth for grinding plant material. A cony has rodentlike incisors, but in elephants and dugongs the upper incisors are modified into tusks. A manatee has neither incisors nor tusks.

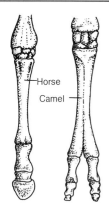

Odd-toed and even-toed animals make up the ungulate group. The horse is an odd-toed ungulate (Perissodactyla), and the camel is an even-toed ungulate (Artiodactyla).

Ungulates

The ungulates group is composed of several unrelated species of mammals that have evolved hoofs. It is divided into two orders—the odd-toed ungulates (Perissodactyla), which comprise about 15 species, including horses, rhinoceroses, and tapirs, and the even-toed ungulates (Artiodactyla), which consist of approximately 170 species, including pigs, deer, camels, llamas, cattle, and hippopotamuses. The characteristics common to all ungulates are related either to the way they walk and run, the way they digest the plant food they eat, or the way they defend themselves. Ungulates are the only horned animals. Those that do not have horns use their long canine teeth or hooves as a means of defense.

General features

All ungulates have large, flattened cheek teeth with good grinding surfaces. Such teeth give them the ability to thoroughly chew the tough plant materials that make up their diet. These cheek teeth, or molars, have high crowns, which can be ground down considerably before they wear out. The cheek teeth have complex ridges composed of dentin, which are next to areas of enamel, and are filled with cement. As the vegetation is ground down, the cement, enamel, and dentin are worn away to different degrees, producing a rough, self-sharpening surface.

Any animal unable to defend itself well against predators needs to be able to run fast—and most ungulates have evolved an efficient type of movement known as unguligrade locomotion. In fast-running animals, the upper bones in the limbs are short, but the lower bones are long, allowing lengthy strides. In addition, the bones in the feet are elongated, and by running on its toes the animal effec-

tively adds a third functional segment to its limb. The toes themselves are lifted until they touch the ground only at the tips, which are protected by strong hoofs. Hoofs, like toenails, are composed of keratin, which makes them light, resilient, and tough.

Because the ungulates lift the backs of their feet to run on their toes, the short side-toes cannot reach the ground. Without a function, these toes have evolved to shrunken, useless appendages in many ungulates or have disappeared altogether. The number of toes on an ungulate determines the order to which it belongs: with the odd-toed perissodactyls, which have one or three digits, or with the even-toed artiodactyls, which have two or four toes.

Ungulates breed usually once a year, with single offspring, although some species breed once every two years. The females come into estrus several times a year and, if not mated, can continue in heat for several months. Their gestation period is 11 to 12 months, and their life span is from 25 to 40 years or more.

Rumination

By far, most ungulates are herbivorous, although some are omnivorous. Unlike most mammals, herbivores can break down cellulose, the chief component of plant cells, and extract nutrition from it. In this method of digestion, called rumination, microorganisms that live in an ungulate's digestive system dissolve the cellulose. Some ungulates ruminate, but they are not considered true ruminants. In these animals, the stomach has three compartments and lacks a rumen.

In true ruminants—giraffes, deer, antelope, sheep, goats, and cattle—the stomach consists of four chambers: the rumen, the reticulum, the omasum, and the abomasum. Food, mixed with saliva, is fermented in the rumen and reticulum, where bacteria and protozoa break it down. To ensure that the food is well broken down, the animal regurgitates it, or brings it

Teeth types vary among the ungulates. The grazing horse (A) has high-crowned cheek teeth with self-sharpening grinding surfaces. The tapir (B), which nibbles at low shrubs, has relatively low-crowned cheek teeth and its nose extends into a short, flexible snout. Among the artiodactyls, the peccaries (C) have tusklike upper canines and a long snout. The camels (D) have spatula-shaped incisors that project forward for grazing.

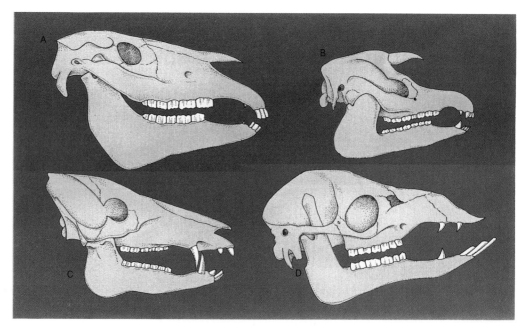

back up, in a mass called the cud and chews it once more. After the food has been chewed and swallowed for the second time, it bypasses the rumen and enters the omasum. Here the mixture is fermented further. The liquid is pressed into the abomasum, where it is subjected to the action of digestive enzymes. The nutrients are absorbed, and the waste is passed out of the body.

Perissodactyls

In the early days of mammalian evolution, many species of odd-toed ungulates existed, but today only three of the original twelve families remain: the horses (Equidae), the tapirs (Tapiridae), and the rhinoceroses (Rhinocerotidae). The tapirs and rhinoceroses are greatly reduced in numbers and, except for the zebra, there would probably be very few horses if they had they not been domesticated. The natural range of wild horses today is Africa and the steppes of Asia.

The perissodactyls have three toes on the hind feet, as in tapirs, or one single toe, as in the equids. They are browsers and grazers, and the structure of their lips facilitates the collection of plant material. Their flat-topped cheek teeth and high-crowned molars enable them to break down coarse vegetable food. They have simple stomachs, a large cecum, and no gall bladder. Horns composed entirely of skin material may be present, and the skin is often very thick, with little hair.

The family Equidae contains only one living genus, *Equus*, which includes horses, asses, and zebras. All species are highly specialized for swift movement and for grazing. Only the third digit remains on the limb. The remains of what were once two other toes grow as bony strips on the canon bone of the horse's legs.

In the wild, Equidae live in migratory herds, as do many herbivores that live on plains. The wild horse is now represented by only one species, the Mongolian wild horse, or Przewalski's horse. They used to roam the steppes of central Asia in large herds but are now extinct in the wild. However, animals born in

zoos will soon be reintroduced into the wild. The wild horse is shorter than the domestic one, with a stiff, erect mane, small ears, a low-slung tail, and a shrill voice.

The ass, native to Africa and Asia, is found most frequently on plains sparsely covered with low shrubs. It was the first animal in the genus to become domesticated.

Zebras, now the most common member of the genus in the wild, are found only in eastern, central, and southern Africa. They were once called horse tigers because of their stripes. The stripes of the plains zebra are wide apart, with shadow stripes in some species. The rare mountain zebra has wide stripes with a transverse gridiron pattern on the lower back. The largest zebra, Grevy's zebra, has narrow stripes.

The tapir, whose natural range is restricted to the Malay Peninsula, Java, Sumatra, and Central and South America, has only one living genus *(Tapirus)* with four species. The

Wart hogs are typical of the hoglike artiodactyls in that their canines form tusks, which they use in defense. The nyala—like many other antelopes—has developed, instead, sharp horns with which it can defend itself.

The well developed canines of the hippopotamuses are used as powerful weapons. These tusks can be seen when the animal opens its mouth. Males use them in fights during the mating season.

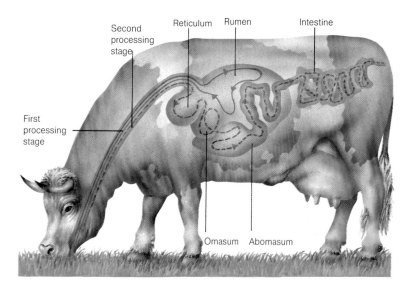

Second processing stage

Reticulum   Rumen   Intestine

First processing stage

Omasum   Abomasum

Cattle are true ruminants. They have a four-chambered stomach composed of a rumen, a reticulum, an omasum, and an abomasum. Food is swallowed along with salivary enzymes and is fermented in the rumen, where the enzymes and bacteria in the digestive juices break it down. It is regurgitated in the form of cud and chewed again. The food is swallowed once more, this time passing into the other chambers, where it undergoes further enzyme action. The nutrients are absorbed, and the waste is passed out of the body.

tapir's forefeet have four toes, whereas the hind feet have three.

The three genera of rhinoceros are found in Africa, Asia, Sumatra, Java, and Borneo. They inhabit savanna and moist wooded areas, often near water. Rhinoceros feed on shrubs, leaves, and fruit, and are mainly nocturnal. These massive animals—they reach 6 to 15 feet (2 to 4.6 meters) in length—have short, pillar-like limbs with three digits. They have thick, sparsely haired skin with characteristic folds and one or two horns on the nose, or nasal plate, composed of solid fibrous keratin. And their hearing and sense of smell are well developed.

Artiodactyls

In the early days of ungulate evolution, artiodactyls were not as common as perissodactyls. But they have become increasingly prominent and today are one of the most successful groups of mammals. They are found throughout the world, except in Australia and New Zealand, where they have been introduced by people.

There are nine families of artiodactyls. Five families are not true ruminants: the pigs and hogs, or Suidae; the peccaries, or Tayassuidae; and the hippopotamuses, or Hippopotamidae; the camels and llamas, or Camelidae; and the chevrotains, or Tragulidae. The rest of the artiodactyls are true ruminants and are grouped into the infraorder Pecora. This group contains the giraffes, or Giraffidae; the deer and their allies, or Cervidae; the cattle, antelope, goats, and sheep, or Bovidae;

and the pronghorn, or Antilocapridae.

The hoglike artiodactyls are in many ways the most primitive of the group. They are still four-toed—although the side-toes are shrunken like dewclaws—and their limbs are not greatly elongated, which means that they cannot move quickly. They are omnivorous and have large, canine tusks. And their two-chambered stomach is not as developed as that of the ruminants.

The most typical of the five genera of wild hogs is the wild boar, from which the domestic hog was probably derived. The hog is typically a nocturnal forest dweller and travels in groups of up to 50 individuals. But some species, such as the warthog, are active during the day and travel singly or in family parties. The males of such species have upper and lower canine tusks that curve upward. They eat roots, plants, birds' eggs, and small mammals. The young number up to 12 per litter.

Peccaries are distantly related to wild hogs. There are three living species of peccaries: the collared peccary, or javelina; the white-lipped peccary; and the tagua, or Chacoan peccary. The tagua lives in the Gran Chaco region of Paraguay, Argentina, and Bolivia. Scientists discovered the tagua in 1975. They had previously thought that this species had become extinct more than 10,000 years ago.

There are two genera and two species of hippopotamuses, both found in Africa. These large, heavy animals are adapted for both aquatic and terrestrial life. They have a broad muzzle that is suitable for taking in large masses of pulpy water plants. Their eyes and ears are situated high up on the skull, en-

The horse is the fastest, strongest runner of the ungulates, because it has the most developed unguligrade limb. As a migratory animal, it is adapted to sustaining speed for long periods.

The zebras are the most common equid in the wild. No two individual zebras are alike in the pattern of their stripes. It is thought that the stripes serve a purpose in social recognition. On the open plain, zebras seek safety in numbers and move in vast herds composed of both males and females. This characteristic contrasts with many of the other plains herd animals, such as antelope, which move in herds of separate sexes.

abling them to function well while almost totally submerged. They are expert swimmers and can stay submerged for more than five minutes. Hippos live in herds of 5 to 30, usually near the water.

The camelids can be split into two groups: camels, which live in the deserts of Africa and Asia, and llamas, which are found in South America. Camels are able to conserve water by reducing evaporation and concentrating their urine, and can lose up to 25 per cent of their body weight by water loss. They also carry a food supply in the form of a hump, which is a large lump of fat that provides energy for the animal if food is scarce.

There are two chief kinds of camels, the Bactrian camel, which has two humps, and the dromedary, which has one. Nearly all camels that exist today are domesticated. Only a few hundred wild Bactrian camels may still wander in remote areas of Mongolia.

The South American camelids, or llamas, live in a variety of habitats, from cool plains to mountains of permanent snow. The llamas are smaller than camels, have no humps, and are covered with a thick coat of wool. Two wild species of South American camelid are the vicuña and the guanaco.

The hoofs have disappeared on all camelids and have been replaced by a nail and a large pad, which enables them to walk on soft or sandy ground. They have two digits on each limb. The camelids are ruminants but have no separation of the omasum and abomasum.

The chevrotains, or mouse deer, are timid forest browsers, not much bigger than a rabbit, found in the tropics of the Eastern Hemi-

The Brazilian lowland tapir is a solitary animal found in damp forests or swamps, and it is nocturnal or active at dusk. It feeds on leaves, shoots, and fruit that it collects with its short, flexible snout.

sphere. They have no horns and use their large, tusklike upper canines as weapons. They resemble deer in the white patterning of their red-brown coat. With a three-chambered stomach, the chevrotains—like the camels—represent a halfway point between the non-ruminants and the true ruminants.

The rest of the artiodactyls, all of which are true ruminants, contain the most successful and numerous of the ungulates. Almost all of them have horns or antlers and a rumen. The side-toes of their limbs have disappeared, leaving two functional digits, although lateral ones may be represented on these animals by imperfect dew-claws. The incisors have been lost from the upper jaw, and the lower teeth

In a transverse gallop, such as this *(below),* a horse's body is supported by at least one limb on the ground, except during one short phase in the cycle.

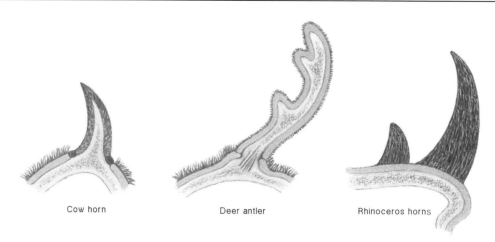

Horns and antlers differ in structure. A cattle horn *(left)*, which is a true horn, has a bony core that is an extension of one of the skull bones. It is covered by a tough material called keratin. A deer antler *(center)* grows out of skull bone but is shed every year. Early in its maturity, an antler is covered by a velvetlike covering of skin. Rhinoceros horns *(right)*, which are not true horns, are not attached to the skull, but they are also not shed. They are made of a fibrous form of keratin.

Cow horn   Deer antler   Rhinoceros horns

bite against the upper gum.

Among these artiodactyls are the giraffes and okapis, browsing animals now found only in tropical Africa. Their long necks contain the usual number of vertebrae for mammals—seven—but each vertebra is greatly elongated. The neck of the okapi is shorter than that of the giraffes. Horns, which are bony knobs that grow continually but slowly, are found in both sexes. Okapis are solitary, but giraffes usually move in loose knit groups or family herds.

The deer, or cervids, which are mainly forest dwellers, are essentially temperate-zone animals, although some are found in southern Asia. The males have antlers, which are bony growths from the skull. They are shed each year and form progressively more branches as the animals mature. The older the cervid, the larger are the antlers.

Cervids are divided into four subfamilies: Moschinae, Muntiacinae, Cervinae, and Odocoileini. Moschinae contains only one genus, the musk deer. Restricted to Asia, this genus is thought to be the most primitive of the cervids and, with the water deer, is the only type with no antlers. The male has an abdominal gland that exudes musk. Because of the demand for musk by the perfume trade, the musk deer has been hunted out of most of its former range.

Muntiacinae, or muntjacs, contains two gen-

era that are also restricted to Asia. They are small deer, and although they have antlers they also have well-developed upper canines. They are solitary animals that are most active at dusk.

The subfamily Cervinae includes the red deer and wapitis, the axis deer, and the fallow deer. They are found in Asia, Europe, and North America. The Odocoileini is the most widely distributed subfamily and includes such species as white-tailed deer; moose and elk; and caribou and reindeer, whose range is the northernmost of all species of cervid.

The bovids are the largest group of ungulates found throughout the world. They include such animals as antelopes, gazelles, and most of the animals that human beings depend on for food, such as cattle, sheep, and goats. There are nine subfamilies: the cowlike Bovinae; the duikers, or Cephalophinae; the reedbuck and their relatives, or Hippotraginae; antelopes and gazelles, or Antilopinae; and the sheep and goats, or Caprinae.

All bovids have bony horns which, unlike antlers, are never shed. The horns, which are firmly attached to the skull, can occur in many different shapes, such as spirals or curves, but they are not branched. Bovids are usually found in herds, and most inhabit grassland, scrubland, or deserts, but goats and sheep are generally found in rocky mountainous or

A stag's antlers are grown and shed every year, each time developing more points, until the animal reaches about its fifteenth year. The antlers are shed in early spring, and six weeks later, new ones develop. By the end of May they are fully grown.

Eighteen months old

First set

Two and a half years old

Second set

Four and a half years old

Fourth set

Fifteen and a half years old

Fifteenth set

Giraffes are the tallest of all animals, the males reaching a height of 18 feet (5.5 meters) and the females 15 feet (4.5 meters). Because rising from a lying position is so difficult for these animals, they usually sleep standing up. They have massive hearts, needed to force the blood up their long necks to the brain. An advantage of their height is that they have little competition for the leaves on high branches on which they feed.

desert areas. They feed by twisting vegetation around the tongue and cutting it with the lower incisors.

Domestic cattle, sheep, and goats have been bred from wild European and Asian species. There are six species of wild sheep found in North Africa, Canada, and mountainous areas of Asia and the Mediterranean. The largest is the argali of the Altai Mountains in Siberia and Mongolia. Males stand 4 feet (1.2 meters) high at the shoulders and have large spiral horns. None of the species of wild sheep has as woolly a coat as the domestic sheep.

Most wild goats live in Asia. One species, the ibex, lives in the mountainous areas of Sudan and Siberia, and in the Alps and Caucasus Mountains. The male ibex has spectacular curved horns that may grow more than 30 inches (76 centimeters) long. Despite its name, the Rocky Mountain goat is not a true goat, but a type of animal called a goat antelope.

Together with rats and hogs, feral goats—goats that have escaped domestic captivity—are among the most destructive of animals because they eat anything in sight. In destroying vegetation, they have been instrumental in the extinction of other animal species, such as some birds.

The pronghorn, native only to North America, is the only remaining species of its family. Herds of about a thousand animals used to be common, but their numbers are greatly reduced, and today they live in herds of about 50 to 100. Like cattle, their horns consist of a bony core that is never shed. Pronghorns do, however, shed the outer covering of their horns. The horns are branched, as are those of the cervids.

Impala gather in large herds, often in groups of hundreds of individuals, which affords them some safety from predators. While a few sentries keep watch, the others can eat in reasonable peace. Impalas can run very fast when threatened and are known to leap as far as 30 feet (9 meters).

Endangered animals

The great auk, a large flightless bird with similar habits and habitats to those of penguins, became extinct in 1844 as a result of overhunting by humans. It can now be seen only as a stuffed museum specimen.

The World Conservation Union has classified more than 700 species of animals worldwide as endangered—that is, faced with extinction unless humans provide direct protection, and another 800 as vulnerable, or threatened. Although some extinction is part of the natural ebb and flow of life on the planet, many human activities contribute to the dwindling numbers of many of the earth's animals.

The existence of life on earth depends on a narrow range of acceptable conditions, and even small variations in such factors as rainfall, temperature range, or chemical pollution levels can lead to the extinction of a variety or a species of animal. And each extinction inevitably has some effect on all the local species, as the ecological balance readjusts to compensate and regain stability.

A larger climatic or other change can have even more dramatic effects. For example, there is strong evidence that the sudden and total disappearance about 65 million years ago of the dinosaurs (then the dominant life form) and many other species resulted from catastrophic climatic changes caused by a collision between the earth and an asteroid. Ice ages, which might have been caused by variations in the sun's activity, had a dramatic effect on plants and animals, leading to the reduction or extinction of some species and the increase or evolution of other better adapted ones.

Effects of one species on another

The gradual evolution of better adapted life forms causes the reduction in numbers and possibly the eventual extinction of less well-adapted species trying to occupy the same ecological niche. In addition, groups of plant and animal species become interdependent in areas of reproduction, habitat, or food requirements. Any external event that affects the numbers or habits of one of these species may therefore have disastrous effects on the survival of the other dependent species. Such interdependence reduces the chances of survival of the species concerned. And disease—particularly the evolution of new virus strains—threatens certain species with extinction while leaving others relatively unaffected.

In each ecological environment one dominant species is likely to evolve, and it has a major influence on the ecological balance, affecting the numbers and survival of other species. The dominant species is most likely to be the predator at the top of the food chain. It will be challenged and replaced as the dominant species only by the evolution of a better adapted species, or by a physical change in the environment.

The activities of humans

The evolution of humans as an intelligent and highly adaptive species has led to a situation in which they have a position of dominance over all other species on earth. This fact, linked with the ability of humans to modify or even totally alter the environment, has placed many species in danger of extinction. Pressures of food production for a rapidly expanding human population have meant that there is probably no natural habitat that is not now threatened.

The threat of humankind to other species is apparent in a number of ways. The development of domesticated strains of plants and animals to improve the efficiency of food production (for example, wheat and domestic cattle) demands the creation of specialized environments that make large areas of habitat unavailable to the native plants and animals. Indeed, people take positive action to exclude the naturally occurring plants and animals from these

Uncontrolled hunting and poaching have long threatened some of the world's large animals with extinction. Throughout the nineteenth century, the North American Bison *(above)*, or buffalo as it was known, was relentlessly hunted so that by the 1880's, its numbers had fallen alarmingly. The animal was brought back from the edge of extinction by careful management. Today, in Africa, the rhinoceros *(right)* is threatened by the wasteful killing of poachers. Poachers kill the rhinoceros only for its valuable horn. The carcass is then left for the vultures and hyenas.

areas with such weapons as weed-killers, insecticides, traps, and fences.

Increased requirements for living space for people remove land from the natural environment as new towns and cities are built. Relocation of waterways and the physical manipulation of the landscape further eliminate natural habitats. In addition, overgrazing of domestic animals can cause soil erosion, reducing previously fertile land to desert, again changing the local ecological balance.

The exploitation of a species for food, clothing, or ornament can reduce its numbers to below that necessary to maintain a population that can survive. Human activities such as farming and industry create pollution by releasing poisonous chemicals into the atmosphere, soil, and water. Species with little tolerance to such pollutants may be threatened. Sometimes catastrophic accidents that release large amounts of oil, toxic gases, or radioactive waste can have immediate and disastrous effects on wildlife. The burning of wood and fossil fuels to produce energy may have long-term effects on the carbon dioxide content of the atmosphere, possibly negatively affecting the climate. Other experiments in climate control might have unpredictable effects on the world's ecology.

The deliberate or accidental introduction of species from different parts of the world into an ecology can have a disastrous effect on local species and may result in the extinction of those that are unable to compete or those that the new species prey upon.

Endangered lower animals

Increasingly, people are becoming aware that various species of birds and mammals are in danger of extinction. But many other animals, such as mollusks, insects, fishes, and reptiles, are also endangered. Colorful tropical butterflies are caught by the thousands so that their wings can be used in jewelry and ornaments. Some species of snails, crabs, and crayfish are caught and eaten as delicacies in restaurants throughout the world. Fishes continue to suffer from the effects of overfishing and pollution of the seas and oceans.

Nearly all such lower animals are part of the food supply of an interdependent chain of higher animals. Eventually, some birds and mammals also may become threatened as their food supply diminishes. And ultimately, the threat may extend to the food supplies of humans themselves.

The greatest of the threats to lower animals is pollution—from the testing of nuclear weapons to the dumping of sewage and industrial waste. For example, the Puerto Rican crested toad, the Virgin Islands boa, and the Aruba Island rattlesnake now exist on only a few islands because humans have claimed

Unique to the Galapagos Islands, the land iguana is being threatened by introduced predators, particularly rats that destroy and eat its eggs and young.

Israel painted frog
Discoglossus nigriventer
Possibly extinct

Italian spade-footed toad
Pelobates fuscus insubricus
Almost extinct

Desert slender salamander
Batrachoseps aridus
250—500

Orange toad
Bufo periglenes
5,000—10,000

Santa Cruz long-toed salamander
Amblystoma macrodactylum croceum
About 10,000

Illinois mud turtle
Kinosternon flavescens spooneri
Fewer than 20

False gavial
Tomistoma schlegelii
About 100

St Croix ground lizard
Ameiva polops
About 150

Watling Island ground iguana
Cyclura rileyi rileyi
200

Indian gavial
Gavialis gangeticus
450

Endangered amphibians *(far left)* include various species of frogs, toads, and salamanders, many of which suffer from the drainage or pollution of the waters in which they breed. Among reptiles *(left)*, turtles and crocodiles are threatened mostly by being hunted for their shells and skins. The numbers given refer to the estimated survivors in the 1990's.

Large land reptiles continue to suffer from the spread of human activities and introduced predatory species. The Galapagos land tortoise *(above, left)* and the Indonesian Komodo dragon *(above, right),* the world's largest lizard, are both endangered.

their natural habitats for tourism and agriculture. Their continued survival is dependent on reintroduction of zoo-born animals and careful management of protected areas.

An instructive example of the threat to lower animals in the sea is provided by the myriad array of fishes and other creatures that depend for their existence on Australia's Great Barrier Reef. Since the 1960's, some parts of the coral reef have died off as a result of pollution and a plague of large, dark, spiny starfish known as the crown-of-thorns. These starfish, often as many as 15 of them on each square yard or meter of reef, prey on coral-forming animals.

The large increase in the number of crown-of-thorns is due in part to the removal from the reef of their chief natural enemies—and population stabilizers—the large marine snails called tritons and helmet shells, whose ornamental shells were much prized by tourists. The sale of these shells has now been halted. If the snail population increases, it will keep the crown-of-thorns population in check, which will allow the coral to revive. Thus, a healthy balance of species will be restored to the reef.

Concentrated fishing and pollution has resulted in shortages in the world's fish popula-

tions. Since the 1940's, the fishing of some species has become so intensive that even common fishes such as flounders, ocean perch, lake herring, and tuna are presently in danger. Often the cause lies not merely in the huge amounts of fish that are caught, but in the fact that, by using nets with a very small mesh, fishers are catching immature fish before they have a chance to breed. Also, the sale of highly valued tropical fish for home aquariums has threatened some of these species.

Most of the major species of endangered lower animals on the land, and in its associated freshwater lakes and rivers, are reptiles. In the Americas, two tortoises, the Mexican gopher and the giant Galapagos land tortoise, are at risk, and the population of the Cuban crocodile has been estimated at fewer than 500 individuals. The short-necked tortoises of western Australia probably total only half that number. And on the small islands off the North Island of New Zealand, the unique, primitive, lizardlike animal called the tuatara—the sole remaining species of a whole order of reptiles that has remained virtually unchanged for 20 million years—is in danger of joining the dinosaurs in extinction.

Endangered birds

Ever since the last dodo disappeared from the island of Mauritius in 1681, about 80 species of birds have become extinct. Today, there are about 210 species and subspecies in danger of extinction. The main threats are to large birds. They are conspicuous and easy to kill, and they breed slowly and thus cannot compensate for hunting or other unnatural mortality. Birds that are confined to islands are particularly at risk because they have small populations and often are not afraid of humans.

Butterflies have long been collected indiscriminately, mainly because they are so decorative. This New Guinea species, the Victoria birdwing, has been hunted almost to extinction.

Various human activities directly threaten bird populations. The most obvious is hunting for food, feathers, or sport. Although a significant threat to birds in many parts of the world, uncontrolled hunting is declining in northern Europe and North America, with more enlightened attitudes about conservation. But controls have come too late for some species, such as the Eskimo curlew. Once seen in huge migratory flocks numbering thousands of birds, the Eskimo curlew has not been sighted for several years and may well be extinct.

The California condor is now extinct in the wild. One of the world's heaviest flying species, with an 8-foot (2.4-meter) wingspan, this majestic vulture was once scarce but widespread over much of North America. The primary reason for its decline is land cultivation—it needs huge hunting areas in which to find animal carcasses—and the small population has been further depleted by illegal shooting. Another example of such a large,

slow-breeding bird is the Steller's, or short-tailed, albatross. Japanese plume traders have slaughtered millions of them, so that the albatross is now reduced to a small colony on Torishima Island, near Japan.

To this day, hunting in areas such as the Mediterranean accounts for many thousands of birds killed. In Italy, for instance, about 100 million songbirds such as robins and thrushes are killed every year. Such pressures can result in local extinctions or near extinctions, as with the threatened European population of the great bustard or the extermination of the Arabian race of the ostrich in the 1960's as a result of elaborate hunting expeditions. An indirect hazard of hunting is that birds eat the waste lead shot scattered over the ground from shotguns. Lead poisoning is a particular threat to swans, geese, and ducks.

Rare species are threatened by unscrupulous egg collectors and taxidermists. Illegal collecting of live birds for the pet trade is also

Many birds face extinction. The Guam Rail *(left),* a flightless bird found only on the island of Guam, became extinct in the wild in 1986, due to the invasion of an alien predator, the brown tree snake. Efforts have been made to breed the rail in captivity with the hopes of reestablishing it in the wild. The bald eagle *(right),* found only in North America, has been considered an endangered species since the 1960's. Hunting and the loss of wilderness regions have caused a great decline in bald eagle populations. Pesticides and industrial wastes in the environment have also interfered with the birds' ability to reproduce.

A   Cuban ivory-billed woodpecker
    Campephilus principalis bairdi
    Fewer than 12

B   Mauritius kestrel
    Falco punctatus
    About 25

C   White-breasted silver eye
    Zosteros alcogularis
    Fewer than 20

D   Madagascar sea eagle
    Haliaeetus vocifcroides
    About 20

E   Pink pigeon
    Nesoenas mayeri
    About 50

F   Chatham Island pigeon
    Hemiphaga novaeseelandiae chathamensis
    Fewer than 25

G   Lord Howe currawong
    Strepera graculina crissalis
    30—50

H   Seychelles magpie robin
    Copsychus seychellarum
    Fewer than 40

I   California condor
    Gymnogyps californianus
    About 100

J   Western tragopan
    Tragopan melanocephalus
    About 50

Of the world's endangered birds, more than 30 species had fallen in numbers to fewer than 50 individuals in the wild by the early 1990's. Even some of those illustrated here have not been sighted for several years and may by now have become extinct. However, many species, such as the pink pigeon, Mauritius kestrel, and the California condor may recover if animals born in zoos and other facilities can be successfully reintroduced into their natural habitat.

a problem. Exotic tropical birds, particularly parrots, suffer most. Huge numbers are captured, and many die crammed in packing cases in aircraft.

Another direct threat occurs when birds are killed because they are thought, often wrongly, to be a pest. The only North American parrot, the Carolina parakeet, once was common, but by 1904 had been hunted to extinction because of its fondness for fruit, especially the citrus fruit in settlers' orchards. Unfortunately, such orchards had replaced much of the birds' original forest habitat.

The main damage to birds, however, is being done indirectly by habitat destruction or alteration and pollution. The most serious threats are to birds of tropical rain forests and wetlands. Logging has destroyed most of the habitat of the Cuban and American races of the ivory-billed woodpecker, for example, and for many years they were feared to be extinct. Although the American population is definitely extinct, in 1986 at least two ivory-billed woodpeckers were sighted in Cuba. However, it is doubtful that these birds represent a viable population. The diminutive Kirtland's warbler struggles to survive in its extremely specialized habitat in central Michigan, helped by careful management and strict protection. There are about 1,000 of these small, yellow-breasted birds, which nest only in areas of forest that have been burned.

Wetland species such as the Japanese crane and the whooping crane are similarly threatened as drainage, pollution, and acid rain from factory emissions destroy their habitats. Oil pollution accounts for the deaths of countless thousands of seabirds and waterfowl.

Other indirect threats to seabirds include fishing nets. Danish fishing boats off Greenland are thought to have killed an estimated 500,000 Brunnich's guillemots in their salmon nets.

Pesticides have had a devastating effect on many birds, such as the peregrine falcon. In the 1950's, the peregrine was almost wiped out in North America and severely reduced in Europe because it ate prey that had been feeding on crops sprayed with DDT and other pesticides. Other birds that have suffered similarly include the American bald eagle, the osprey, various species of pelicans, and the bald ibis. However, with the ban on DDT in the 1960's, these species have rebounded and the status of bald eagles has improved from endangered to threatened.

One of the worst of all indirect threats comes from animals introduced by humans, especially on islands where there are no indigenous mammalian predators. Cats, rats, dogs, and hogs have helped wipe out many species. Rats, for instance, have caused the extinction of at least nine species of flightless island rails, and cats introduced in 1931 to Herokapare Island, New Zealand, have wiped out six species and reduced the total bird population from an estimated 400,000 to a few thousand.

Endangered mammals

Almost every year for the past 80 years, at least one species of mammal has become extinct. Of the more than 4,500 or so living species of mammals, 475 are currently listed as endangered and are given protection by international law from hunting and exploitation.

Conspicuous and often posing physical competition or danger to humans, mammals are on the front line of the fight for survival. Humans have considerable commercial interest in mammals and depend on them for a variety of purposes. Whales are killed for their oil and meat, monkeys are snatched from forests for medical research, leopards and cheetahs are killed for sport and the profit from the sale of their fur. Other mammals—the giant panda and tiger, for example—are threatened by the destruction of their natural environments caused by concentrated exploitation of valued natural resources or through conversion of forest land for agriculture. The discharge of chemicals and liquid waste into lakes and rivers is another hazard.

Special parts of some mammals are sought by the fashion and jewelry trade. Rhinoceros horn is a highly prized ingredient in East Asian health potions and is carved into dagger handles in several African and Arabian countries.

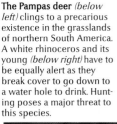

The Pampas deer (below left) clings to a precarious existence in the grasslands of northern South America. A white rhinoceros and its young (below right) have to be equally alert as they break cover to go down to a water hole to drink. Hunting poses a major threat to this species.

Cuban solenodon
Atopogale cubana
Fewer than 20

Eastern cougar
Felis concolor concolor
About 20

Greater bilby
Macrotis lagotis
About 20

Northern hairy-nosed
wombat
Lasiorhinus krefftii
Fewer than 40

Southern bearded saki
Chiropotes satanus satanus
Fewer than 50

Vancouver Island marmot
Marmota vancouverensis
Fewer than 100

Golden marmoset
Leontideus rosalia
Fewer than 100

Buff-headed marmoset
Callithrix flaviceps
About 100

Lower Californian
pronghorn
*Antilocapra americana
peninsularis*
About 100

Volcano rabbit
Romerolagus diazi
Fewer than 200

**The most endangered
mammals** range in size
from the Cuban species of
the shrewlike solenodon to
the cougar and pronghorn
of the United
States. Once the numbers
of a particular species fall
below about 50, it is ex-
tremely doubtful whether a
viable breeding population
can be maintained without
special protective measures
being taken. The numbers
given here reflect estimates
of the early 1990's.

Ambergris, a substance produced in the bodies of sperm whales, is used as a base for per-
fumes. Elephant and walrus ivory are prized
for use in a variety of decorative objects and
jewelry and was once used for piano keys.
Crocodile skin is also in demand for shoes,
belts, and other leather goods. Even when animals are protected by law, poachers will go to
great lengths to obtain them.

The whale is a potent symbol of endangered
aquatic mammals, and of the need for international management and regulation. In the
1800's, the herds of whales that gathered in
the plankton-rich waters of the Antarctic were
extensively hunted and killed. Modern intensive hunting has left many species at the edge
of extinction. A hundred years ago, the blue
whale—the world's largest mammal—num-
bered nearly 250,000. By 1965, a mere 2,000
were left, and some experts believe that the
animal is now beyond saving. International
concern over the blue whale caused Antarctic
whalers to "mass hunt" smaller species. As a
result, all of the eight species of whales are
now in danger. The gray whale of the Pacific
was near extinction earlier in this century.
However, a ban on whaling has brought the
population back to a nonendangered status.
The humpback whale has been reduced to an
estimated 1,000. Fin, sperm, sei, and minke
whales are also threatened.

Other sea mammals are also protected. Re-
strictions on hunting have saved the polar
bear and various species of seals. The Arctic
fox is also given limited protection.

The fate of the European bison, the largest
of Europe's mammals, was determined by a
high demand for the land over which they
roamed. By the beginning of the twentieth
century, the bison was almost extinct, as were
the wolves and bears that occupied the same
habitats. National parks throughout Europe
now offer sanctuary not only to the bison but
also to elk, deer, beaver, otter, and other

mammals. Bears are protected in special
reserves.

Once found over most of the northern half
of the world, the wolf is now becoming scarce
in many countries. It is rare in western Europe
and extinct in 11 countries, killed for a variety
of reasons, most often because farmers believed them to be a threat to their livestock.
Once common in Europe and northern Asia,
the European lynx is now scarce, although
government protection is helping stabilize the
population and there are plans to reintroduce
animals bred in zoos.

The colonization of Africa over the last two
centuries led to the over-hunting of many
large mammals, and many species disappeared from their former ranges. Many populations declined as well, as forests were cut
for agriculture and ranching. During the

Prosimian primates are
among the most highly
evolved animals on the endangered list, particularly
those confined to islands,
where they cannot escape
from intensive hunting or
introduced predators. The
sifaka is native to Madagascar, where some of the animals are protected on
reserves.

1880's, for example, nearly 80,000 elephants were killed for their ivory. But this total paled in comparison to the slaughter in the 1980's and early 1990's. Today, there is an international ban on trade in elephant ivory, and elephants are protected on many special reserves.

Many of Africa's familiar mammals are now seriously at risk. A list of threatened African mammals includes the lemurs of Madagascar, the okapi, the mountain gorilla, Grevy's zebra, the eland, and the black rhinoceros. These animals have been given complete protection, and the giraffe, chimpanzee, and some of the big cats—especially the cheetah—are guaranteed a limited safety. Poaching, however, still occurs in the large game reserves. The high price of rhino horn has driven the black rhino to a precariously small wild population of less than 3,000 individuals.

Animals of the Americas also face extinction, including the polar bear of northern North America, the brown grizzly bear of the Rocky Mountains, and the spectacled bear of the Andes. Other endangered American mammals include the black-footed ferret, Mexican pronghorn, mountain and Baird's tapirs, Mexican wolf, pampas deer, chinchilla, and vicuña.

In Australia, the unique native marsupials are found nowhere else in the world, yet these same animals are being threatened by the spread of mechanized agriculture and the search for minerals. One mammal, the striped, wolflike Tasmanian tiger (also known as thylacine) is now thought by scientists to be extinct.

Conservation successes

Despite the number of animal species that have become extinct, the conservation movement has had many successes. One achievement has been the establishment of national parks and reserves. An international conference in Paris in 1968 established that such parks should be at least about 1,920 acres (800 hectares) in extent, with strict prohibitions on mining, cultivation, stock raising, hunting, and fishing. There are now more than 1,400 reserves in the world, covering over 3 billion acres (1.25 billion hectares).

Many national parks in Africa offer protected sanctuary to rhinoceroses and other animals. Kruger Park in South Africa covers nearly 2.5 million acres (1 million hectares) and contains many different habitats, ranging from humid tropical forest to near desert, hot savanna to bamboo forests. Apart from animals such as the wildebeest, Kruger Park harbors groups of other animals, such as zorillas, that would otherwise be extinct.

European national parks and wilderness guarantee the conservation of several varieties of eagles and owls, the peregrine falcon, the rose flamingo, and other birds. The Camargue, a swampy wilderness in France, is one of many wetlands that has preserved a great number of the 150 species of migrating water birds. In Africa, similar sanctuaries have conserved the secretary bird, ostrich, maribou stork, pelican, and many small birds that migrate the length of the continent.

Australia has been very aggressive in protecting its native species. Special efforts are being made to preserve the koala, the platypus, and the echidna. More than 122 million acres (49 million hectares) of terrestrial Australian territory now lie within the boundaries of a number of national parks. Special task forces are charged with protecting and rehabilitating endangered species.

International maritime treaties have also been instrumental in preserving many species of the sea, including whales, fish, dolphins, and seals. CITES, the Convention on the Trade in Endangered Species, regulates trade in species between countries.

Conservation success stories include the polar bear (left), which was saved from extinction by international efforts to ban or severely restrict the hunting of the animals, and the Père David's deer (right), a Chinese animal saved from extinction in Britain. The deer were bred from a small group that had survived on the grounds of an English country house and were introduced into zoos and animal parks throughout the world.

The scimitar-horned oryx was in danger of extinction until it was bred in captivity. Here a pair of these graceful animals feed contentedly in the protection of a European wildlife park, far away from their native Africa. Its close relative, the Arabian oryx, became extinct in the wild. However, there are now more than 300 Arabian oryxes living in the wild as a result of successful reintroductions of oryxes that were born in zoos and released back into the wild.

One particular animal conservation success story involves the preservation of the polar bear. In 1955, the International Union for the Conservation of Nature recommended that all Arctic countries should curb the hunting of bears. In 1965, the Soviet Union banned the killing of polar bears and established a special reserve for them on Wrangle Island. Norway has set up a similar sanctuary. In Canada, only licensed hunters and Inuit, sometimes called Eskimos, are permitted to hunt bears. Despite safari hunts outside territorial waters, the population of the animals has increased to the point that it is no longer classed as endangered.

The preservation of the oryx is another notable conservation success story. By 1960, there were less than 100 Arabian oryxes. Hunters using machine guns and jeeps eventually drove this species to extinction in the wild. However, three oryxes that had been captured in 1962 were bred successfully with other oryxes in captivity, thus saving the oryx from complete extinction. In the late 1980's and early 1990's, an international breeding effort among zoos in the United States and other countries returned the oryx to its natural habitat in Saudi Arabia and Oman.

Captive breeding programs also have been instrumental in conservation efforts on behalf of many other species, including black-footed ferrets, Mexican wolves, white rhinoceroses, golden lion tamarins, American and European bison, timber or gray wolves, swift foxes, and Przewalski's horses.

In spite of these efforts, there are many endangered species whose populations are already so small that, even though some individuals are protected by reserves or in zoos, their future remains in doubt. For example, only about 60 Micronesian kingfishers remain, and all are in zoos. And less than 20 Hawaiian crows remain in the world in a captive facility in the wild. Many other species also exist only in zoos or at other captive breeding facilities. However, captive breeding is only one potential solution in the war against extinction. Efforts to preserve animal habitats must continue and must expand if many more species are to stave off extinction.

Conservation measures have succeeded in increasing the numbers of two northern mammals: the Alaskan, or northern, fur seal *(left)* and the once-rare saiga antelope of the Asian steppes *(right)*. Protected by law, the seal population is currently multiplying. And efforts on behalf of the saiga have been so successful that permits for limited hunting are now available.

Animal taxonomy

The Animal Kingdom

Kingdom	Phylum	Subphylum	Class	
Metazoa	Porifera		Calcaria	Hexactinellida
			Demospongiae	Sclerospongiae
	Cnidaria		Hydrozoa	Scyphozoa
			Anthozoa	
	Ctenophora		Tentaculata	Nuda
	Platyhelminthes		Turbellaria	Trematoda
			Cestoda	
	Entoprocta			
	Rhyncocoela		Anopla	Enopla
	Gnathostomulida			
	Gastrotricha			
	Rotifera		Digononta	Monogonta
	Echinorhyncha			
	Nematoda		Aphasmida	Phasmida
	Nematomorpha		Gordioidea	Nectonematoidea
	Acanthocephala			
	Annelida		Polychaeta	Oligochaeta
			Hirudinea	
	Echinodermata		Asteroidea	Ophiuroidea
			Echinoidea	Holothuroidea
			Crinoidea	
	Mollusca		Gastropoda	Bivalvia
			Cephalopoda	Monoplacophora
			Polyplacophora	Aplacophora
			Scaphopoda	
	Arthropoda	Chelicerata	Merostomata	Pycnogonida
			Arachnida	
		Crustacea	Cephalocarida	Branchiopoda
			Ostracoda	Mystacocarida
			Copepoda	Branchiura
			Cirripedia	Malacostracas
		Uniramia	Insecta	Chilopoda
			Diplopoda	Pauropoda
	Tardigrada			
	Linguatulida			
	Echiura			
	Bryozoa			
	Priapulida			
	Phoronida			
	Brachiopoda			
	Sipunculidea			
	Pogonophora			
	Chaetognatha			
	Chordata	Hemichordata	Pterobranchia	Enteropneusta
		Urochordata		
		Cephalochordata		
		Vertebrata	Agnatha	Elasmobranchiomorphi
			Osteichthyes	Amphibia
			Reptilia	Aves
			Mammalia	

Largest Animal Classes

Class (or Subclass)	Superorder (or Infraclass)	Order		
Arachnida		Scorpiones	Pseudoscorpiones	Solifugae
		Palpigradi	Uropygi	Schizomida
		Amblypygi	Araneae	Ricinuclei
		Opiliones	Acarina	
Malacostraca				
(Phyllocarida)		Leptostraca		
(Eumalacostraca)	Syncarida	Anaspidacea	Bathynellacea	
	Hoplocarida	Stomatopoda		
	Eucarida	Euphausiacea	Decapoda	
	Peracarida	Mysidacea	Cumacea	Tanaidacea
		Thermosbaenacea	Spelaeogriphacea	Isopoda
		Amphipoda		
Insecta				
(Apterygota)		Protura	Thysanura	Collembola
(Pterygota)		Ephemoroptera	Odonata	Orthoptera
		Isoptera	Plecoptera	Dermaptera
		Embioptera	Psocoptera	Zoraptera
		Mallophaga	Anoplura	Thysanoptera
		Hemiptera	Homoptera	Neuroptera
		Coleoptera	Strepsiptera	Mecoptera
		Trichoptera	Lepidoptera	Diptera
		Hymenoptera	Siphonoptera	
Osteichthyes				
(Actinopterygii)	Chondrostei	Acipenseriformes	Polypteriformes	
	Holostei	Lepisosteiformes	Amiiformes	
	Teleostei	Elopiformes	Anguilliformes	Notacanthiformes
		Clupeiformes	Osteoglossiformes	Mormyriformes
		Salmoniformes	Cetomimiformes	Gornorynchiformes
		Ctenothrissiformes	Cypriniformes	Siluriformes
		Percopsiformes	Batrachoidiformes	Gobiesociformes
		Lophiiformes	Gadiformes	Atheriniformes
		Beryciformes	Zeiformes	Lampridiformes
		Gasterosteiformes	Channiformes	Synbranchiformes
		Scorpaeniformes	Dactylopteriformes	Pegasiformes
		Perciformes	Mastocembeliformes	Pleuronectiformes
		Tetradontiformes		
(Sarcopterygii)		Crossopterygii	Dipnoi	
Aves				
(Neornithes)	Neognathae	Struthioniformes	Casuariiformes	Apterygiformes
		Rheiformes	Tinamiformes	Sphenisciformes
		Gaviiformes	Podicipediformes	Procellariiformes
		Pelecaniformes	Ciconiiformes	Anseriformes
		Falconiformes	Galliformes	Gruiformes
		Charadriiformes	Columbiformes	Psittaciformes
		Cuculiformes	Strigiformes	Caprimulgiformes
		Apodiformes	Coliiformes	Trogoniformes
		Coraciiformes	Piciformes	Passeriformes
Mammalia				
(Prototheria)		Monotremata		
(Theria)	(Metatheria)	Marsupalia		
	(Eutheria)	Insectivora	Edentata	Pholidota	Chiroptera
		Primates	Rodentia	Lagomorpha	Cetacea
		Carnivora	Tubulidentata	Hyracoidea	Proboscidea
		Sirenia	Perissodactyla	Artiodactyla	

Glossary

In the following glossary, small capital letters (e.g., NUCLEUS) indicate terms that have their own entries in the glossary.

A

allantois A membrane in the EMBRYO of TETRAPODS. It carries a large number of blood vessels that connect the blood system of the mother's PLACENTA with the embryo and, in reptiles and birds, allows gas exchange between the SHELL and the embryo.

allele A GENE with a pair member, both occupying the same loci on homologous CHROMOSOMES. They pair during MEIOSIS and can MUTATE one to the other.

alteration of generations Found in lower plants and some animals, especially cnidarians, it involves organisms of distinctly different body form, such as POLYPS and medusas, each of which gives rise to the other.

alula Also known as a "bastard wing," this small bunch of feathers is attached to the thumb-bone of a bird's wing. At low flight speeds it can be raised to form a "slot" that controls air flow over the upper surface of the wing, reducing the stalling speed.

ametabolous Describes those insects that hatch from eggs as miniature replicas of their wingless parents and simply grow in size between each molt, without METAMORPHOSIS. The process occurs in the subclass Apterygota, a group of primitive wingless insects, such as silver fish.

amino acid A member of a group of organic acids containing an amino group and a component of DNA molecules in the CELL NUCLEUS. About 25 amino acids are known to be found in PROTEINS. Amino acids are the building blocks of proteins, combining in different orders to form them.

amnion A fluid-filled sac and the innermost membrane in which the EMBRYOS of TETRAPODS develop.

asexual reproduction The formation of a new individual from a single parent. This may be by PARTHENOGENESIS, by budding, or by fragmentation.

B

bacterium A microscopic uni- or multicellular organism. Bacteria vary in shape (the basis of their classification) and motility. They are generally considered to be plantlike although they lack CHLOROPHYLL. They feed on plant and animal tissues (as well as inorganic matter), decomposing it and thus releasing in water and soil the nutrients which support larger and more complex organisms.

baleen Plates of horny material hanging from the palate of toothless whales. The inner edge is fringed and forms a filter. Seawater drawn into the open mouth is squirted out through the baleen, food contained in the water is kept behind.

bastard wing See ALULA.

bilateral symmetry The symmetry of most active animals with an elongate body. At one end is a head in which the main sense organs and BRAIN are assembled; behind it the body carries paired organs such as limbs, which are mirror images of each other.

bilharzia Also known as schistosomiasis, this parasitic disease, widespread in the tropics, is carried by flukes of the genus *Schistosoma*. It is transmitted in water when the LARVA of the fluke burrow through the skin. Like most parasitic diseases, it is debilitating and may be fatal.

binary fission Reproduction of a CELL by splitting into two equal parts, common in many single-celled organisms.

binomial classification The system of classifying plants and animals by which each SPECIES is given a unique combination of two names: the genus, or group name, and the species name. The genus name always begins with a capital letter, the species with a small letter; e.g., *Panthera leo,* the lion.

bone Vertebrate skeletal material, formed largely from COLLAGEN, phosphates, and calcium salts. It is hard, but flexible to some extent.

brain The forepart of the main nervous system in BILATERALLY SYMMETRICAL animals. It is housed in the head, close to the principal sensory organs, such as the EYES, ears, and nose. To a greater or lesser extent, the brain coordinates the activities of the entire body.

C

carnassial In carnivores, the first lower molar tooth and the last upper premolar teeth. Together they form large shearing blades and are used for slicing flesh.

cartilage A skeletal tissue in vertebrates containing COLLAGEN fibers in a matrix composed largely of carbohydrates. It is softer and more flexible than BONE.

cell The basic unit of organic tissues. Cells are bounded by a cell wall or membrane and contain a NUCLEUS, which is derived by division from a previous cell nucleus and holds the CHROMOSOMES and CYTOPLASM. Some organisms are unicellular but most comprise many specialized cells.

cellulose A complex carbohydrate material, with a fibrous structure that makes the cell walls of plants rigid.

cephalothorax The fused head and forebody found in arachnids and some crustaceans.

chiasmata The crossing-over or exchange of genetic material between CHROMOSOMES during MEIOSIS. It can lead to the production of new varieties within a SPECIES.

chitin The tough, fibrous EXOSKELETON of arthropods. It is relatively inelastic and therefore dictates the jointed limb and body structure of arthropods and their system of growth by molting.

chlorophyll A green pigment found in most plants, contained in the CHLOROPLASTS. It absorbs light from the sun and by photosynthesis converts it into chemical energy and uses it to produce carbohydrates (sugars and starches).

chloroplasts Small, pigment-containing bodies found in plant CELLS. They contain CHLOROPHYLL as well as other pigments that give plants their color.

choanocyte A collar CELL found in some sponges, in which a flagellum arises from a PROTOPLASMIC collar. The beating of the flagella moves water containing food and oxygen through the sponge's body.

chromatid One of two long filaments of GENETIC material on a duplicated CHROMOSOME, visible during MITOSIS and MEIOSIS.

chromatin The nucleoprotein material of which CHROMOSOMES are made.

chromatophore A pigment-containing cell found in many animals, such as squids, chameleons, and flatfish. The pigment level can be altered rapidly, which enables the animal to change color.

chromosome Found in the NUCLEUS of every plant and animal CELL, chromosomes are composed of CHROMATIN, and carry the GENES, which are the units of inheritance. The number of chromosomes varies in different SPECIES, but in all, the sex cells (GAMETES) contain half of the normal complement. In fertilization, the chromosomes from the male and female gametes come together to give the new individual its complete inheritance pattern, received in part from each parent.

clavicle The collarbone found in vertebrates. The clavicles have been lost in many mammals, but are found in birds as a single, fused element called the furcula, or wishbone.

clitellum A saddlelike structure found in some sexually mature annelids. It produces a mucous sheath around mating animals and binds the cocoon round the fertilized eggs.

cloaca The end of the gut in vertebrates (except for non-marsupial mammals) into which the digestive and reproductive tracts open.

cnidocil The trigger hair of a NEMATOCYST, or stinging cell.

coelenteron The body cavity of cnidarians. It has one opening only, through which food enters and waste products and sperm or eggs are ejected.

coelom The main body cavity of many complex animals.

collagen A PROTEIN which forms intercellular fibers in the skeletal tissues of vertebrates. Collagen fibers have great tensile strength, but little elasticity.

commensalism The association of two different organisms where mutual advantage may result, although the partners may also be capable of independent existence.

compound eye Found in crustaceans and insects, compound eyes consist of a number of separate lens systems (more than 10,000 in each eye in some insects), each of which accepts information from a very small part of the surroundings. Compound eyes lack visual acuity compared with cephalopod or vertebrate EYES.

conjugation A process of SEXUAL REPRODUCTION that involves the exchange of nuclear material during partial fusing of two individual ciliate protozoan cells.

coracoid A strutting bone in the PECTORAL GIRDLES of primitive land vertebrates which lies between the sternum and the outer ends of the CLAVICLES. Today coracoids are only present in birds.

cornea The clear, tough, protective layer of connective tissue that covers the lens of the EYE in vertebrates.

corona The ciliated "crown" of rotifers. The activity of the CILIA is the means of locomotion and food collection in these creatures.

crop A thin-walled, distendable sac in the esophagus of many birds, which functions as a food store.

cypris The fully developed LARVA of a barnacle, when it is ready to settle on a surface.

cytoplasm The watery substance that surrounds the NUCLEUS of a CELL and which contains bodies such as mitochondria and Golgi bodies. Together with the nucleus they form the PROTOPLASM of the cell.

D

deoxyribonucleic acid See DNA.

dewclaws The toes of a DIGITIGRADE or UNGULIGRADE animal that do not normally touch the ground. In carnivores, such as dogs, the first toe of the forelimb forms a dewclaw; in many deer, the second and fourth digits of all feet are dewclaws.

diaphragm A sheet of tissue that separates the chest cavity from the abdomen, found in mammals. It can be arched or flattened to improve the efficiency of breathing.

digitigrade The method of TETRAPOD locomotion in which an animal walks on its toes, thus increasing its stride by the length of the hand or foot bones. Dogs and cats are digitigrade.

dioecious A term applied to plants and animals that have separate sexes.

displacement activity An inappropriate form of behavior performed as a result of conflicting drives. It usually appears as an exaggerated form of normal activity.

DNA (deoxyribonucleic acid) A compound found in plant and animal CELLS, formed from nitrogenous bases arranged in a double helical form and capable of self-replication. It carries GENETIC information in the arrangement of these bases. Each sequence of three bases forms the code for one AMINO ACID.

E

ecdysis The shedding of the EXOSKELETON in arthropods, necessary in this group to allow growth. Under HORMONAL control, the body covering is split off, allowing a new one, already produced beneath the old, to expand and then harden.

echolocation A method of short-distance navigation and sometimes food-finding, used by some nocturnal and aquatic animals. Short bursts of sound are produced by the animal which, if they encounter an obstruction, return an echo which is detected by the animal. Bats and whales echolocate using ULTRASONIC sound.

eclipse period A dull-colored, post-mating plumage phase in many birds. In ducks and geese, this phase is accompanied by a simultaneous molt of the flight feathers, so that they are temporarily flightless.

elytron The thickened and stiffened forewing in beetles.

embryo The pre-birth or pre-hatching stage of a developing animal.

endoskeleton The body support that lies within the muscular structure. It is best developed in vertebrates as a SKELETON, but is also seen in some other groups.

enzymes Proteinaceous substances that are produced by living CELLS, vital to the processes of METABOLISM.

epidermis The outermost layer of CELLS of plants and animals. In vertebrates, it is several cells thick and in land-living forms, the outer layers are often dead and horny. In invertebrates, it is one cell thick and often secretes a CUTICULAR protection.

estivation Dormancy during hot, dry seasons.

estrus The period of greatest sexual receptiveness in a female mammal, normally coinciding with ovulation.

exoskeleton The bodily support that lies outside the muscular structure. It is best developed in arthropods and mollusks.

eye An organ for light-reception, varying from a simple

cluster of light-sensitive CELLS, such as the OCELLUS or the COMPOUND EYE, to a complex system with one or more light-gathering lenses focusing on special cells sensitive to light intensity and color.

F

fetus A mammalian EMBRYO in the later stages of its development.

filter-feeding A method of feeding used by many marine invertebrates in which minute organisms are removed from large quantities of water that is strained through GILLS or some other sifting mechanism.

food vacuole A fluid-filled space found in single-celled animals, created when food is engulfed. The fluid contains ENZYMES which digest the food. The waste flows out from the vacuole into the CELL CYTOPLASM.

frenal hooks Tiny hooks along the edges between the two wings of hymenopterans (bees and wasps, etc.), which hold the two wings on each side of the body together so that they beat as a single unit.

furcula *See* CLAVICLE.

G

gamete A reproductive CELL with half the number of CHROMOSOMES normal for the SPECIES (haploid). It is either a mobile male cell (sperm) with reduced CYTOPLASM, or an immobile female cell (ovum) with hugely increased cytoplasm. They fuse during fertilization when a diploid cell (ZYGOTE), with a full complement of chromosomes, is formed.

ganglion A small knot of nerve tissue from which nerve cords arise. In invertebrates, it may form the BRAIN. In vertebrates, ganglia occur in the peripheral nervous system and as some of the nerve connections of the brain.

gastrulation The movement of CELLS in the early stages of the development of an EMBRYO to the positions in which they will form the internal organs of the animal.

gemmule A bundle of CELLS that are capable of developing without fertilization into a new organism. They may be a means for an individual, the major part of whose body dies in adverse circumstances, to survive until the environment is suitable for further activity.

gene The basic unit of heredity, formed from a sequence of bases on a DNA chain. Each gene has a definite position of the CHROMOSOME and may occur as an ALLELE.

genotype The GENETIC (hereditary) constitution of an individual.

genus See BINOMIAL CLASSIFICATION.

gestation The period of development between conception and birth in mammals.

gill The RESPIRATORY organ of aquatic animals. Gills are usually projections from the body wall or gut and are often complex in shape, so that they offer the maximum surface area for gas exchange between the blood which flows through them and the water which flows around them.

gill-book Found in some aquatic arachnids, these plates of tissue bear a resemblance to the pages of a book; they are set in a cavity where they can be bathed in water from which they extract oxygen and into which they release carbon dioxide.

gizzard The muscular stomach of birds and some other animals, which breaks up food prior to digestion, sometimes with the aid of swallowed stones.

guard hairs The thick hairs which overlie the fine underfur in many mammals. In cold or wet conditions, air is trapped close to the skin, and the guard hairs—which are often greasy and waterproof—clamp down and prevent the escape of this warm blanket of air.

H

heart The strong, muscular pump that drives blood through animals. In mammals and birds, it has a com-

plex of chambers which separate out deoxygenated blood and send it to the lungs, from which it returns oxygenated to supply the body with its needs. In lower vertebrates, the oxygenated and deoxygenated blood is mixed and in invertebrates, it plays little or no part in the distribution of oxygen throughout the body.

hemimetabolous Describes insects with an incomplete METAMORPHOSIS. When they hatch, they resemble the parents, although they do not develop functional wings until mature. They grow by a series of molts.

hemocoele A closed body cavity containing blood. It is well-developed in arthropods and mollusks. Unlike the true COELOM, it never contains reproductive cells.

hemocyanin A RESPIRATORY pigment containing copper, found in the blood of most arthropods and mollusks. It gives the blood of these animals its green color.

hemoglobin An iron-containing RESPIRATORY pigment found in the blood of most vertebrates and some invertebrates.

hermaphrodite An animal possessing both male and female sexual organs.

hibernation A state of torpor which some animals enter when the surrounding temperature drops below a certain critical point. Their temperature, heartbeat, and breathing rates fall dramatically and they use very little energy in staying alive.

holometabolous Describes insects that undergo a complete METAMORPHOSIS from an egg to a LARVA, a PUPA, and finally an adult.

homeothermal Describes animals that are "warm-blooded," with an internal thermostat that keeps their body temperature within very narrow limits, independent from the surrounding temperature.

hormones Substances produced by glands and released into the bloodstream where they have an important role in body functions, such as growth, reproduction, and digestion.

host An individual on which a PARASITE feeds.

hybrid The result of successful cross-fertilization between members of different SPECIES. Hybrids among animals are rare in nature and are almost always sterile.

hydrostatic skeleton The body support of some invertebrates, which depends on the movement of liquid inside the cells to maintain body shape or to change position.

hyoid A system of small bones in the base of the tongue in TETRAPODS, derived from a GILL arch in fish ancestors. In some SPECIES, they have become enlarged to support a long insect-catching tongue, as in woodpeckers, or a resonating voice box, as in howler monkeys.

hypothalamus A part of the BRAIN of vertebrates, which is thought to contain, among other things, the mechanism of body temperature control.

I

iliac crest The expanded upper part of the ilium, which is part of the PELVIC GIRDLE of TETRAPODS. It is particularly important in humans as the area of attachment of the muscles which allow them to stand and walk upright.

imprinting The process by which many animals learn to recognize their own kind. It occurs at an early stage of development as a response to a narrow range of stimuli to which the young individual is normally exposed.

incubation The period between the laying and hatching of certain eggs, when they are kept warm, either by the sun, rotting vegetation, or the warmth of the parent's body.

instinct Nonlearned or innate behavior patterns, often very complex, usually triggered by specific stimuli.

K

keratin A tough, fibrous PROTEIN material found in TETRAPODS. It forms the outer layer of the skin, and hair, hoofs and horny structures in mammals, feathers in birds, and scales in reptiles.

L

larva The young stage of many OVIPAROUS animals that have a distinctly different body form from the adult, usually related to a specialized way of life. For example, in insects the larvae may be the feeding stage, in many crustaceans, the dispersal phase. At the end of the larval life, they rapidly METAMORPHOSE to the adult state.

lateral-line system A series of sense organs on the flanks and sides of the head in agnatha, fishes, and amphibians. Each nerve ending, housed in a bony pit, is sensitive to pressure changes, so that the animals can detect, through water movement, the presence of potential enemies or prey.

lung The RESPIRATORY organ of land-living vertebrates and some mollusks. Air is drawn into a cavity, the skin of which is well supplied with blood vessels, so that gas exchange can take place there.

lung-books The breathing organs of some arachnids. They consist of a series of thin flaps of tissue, like the pages of a book, projecting into a cavity in the body wall. Blood flows through them and gas exchange takes place at their surface.

M

macronucleus A large NUCLEUS which divides MEIOTICALLY and disappears during CONJUGATION in Ciliophora.

Malpighian tubules Excretory glands found in the hindgut of insects, arachnids, and some other types of arthropods.

mammary glands MILK-producing glands of female mammals. Their state usually varies with the ESTRUS cycle and their growth and milk production is controlled by various HORMONES.

mandible The lower jaw of vertebrates, and in some invertebrates (such as insects) the major pair of crushing mouthparts.

maxilla In vertebrates, this forms part of the upper jaw; in invertebrates, it forms one of the pairs of mouthparts that lies behind the MANDIBLES.

meiosis Two stages of CELL division that result in the formation of GAMETES. (See also MITOSIS.)

mesogloea A jellylike substance found between the ectoderm and endoderm (the outer and inner tissue layers) of cnidarians.

metabolism Life processes which involve both the breakdown (catabolism) of organic compounds to liberate energy for various activities, and also the build-up (anabolism) from simple materials of the complex tissues of an organism.

metamorphosis The change that occurs between the immature, LARVAL stage of an animal's life and its mature, reproductive phase. Complete metamorphosis involves a major change in form and takes place within the shelter of a PUPAL case. Incomplete metamorphosis is more gradual until wing growth and sexual development are accomplished.

micronucleus A small nucleus found in Ciliophora which divides MITOTICALLY and provides the nuclear material for CONJUGATION. After conjugation, the MACRONUCLEUS is reformed from the ZYGOTIC (micronuclear) material.

milk A substance produced in the MAMMARY GLANDS of female mammals. It contains fats, sugars, and PROTEINS and is used to feed the young.

mitosis Normal CELL division in which CHROMOSOME material is doubled, so that daughter cells carry the same GENETIC information as the parent cells.

mutation A spontaneous, irreversible GENETIC change. It normally occurs during CELL division and can affect any cells of the body. Some mutations are harmful

and kill the organism, but those that are not lead to variation in the SPECIES.

myoglobin An oxygen-carrying pigment contained in the muscles of mammals, especially those such as seals and whales, which hold their breath for long periods.

N

nectar guides Brightly colored lines or patches on petals that lead insects toward the nectar which usually lies at the center of the flower. In their search, the insects brush against pollen, which they carry to other flowers to effect pollination.

nematocyst The stinging cell of a cnidarian. Thickly scattered on the tentacles, nematocysts consist of a poison cell in which a long, barbed, hollow tube lies coiled, and above which lies a CNIDOCIL, or hair. When this hair is triggered, the hollow tube shoots out and cuts the skin of the prey, allowing the poison to affect the nervous system.

nidifugous A term that describes young birds that leave the parental nest shortly after hatching, such as ducks and game birds.

notochord An internal rod, made of a stiff substance, found in the LARVAE or EMBRYOS of vertebrates. It lies beneath the nerve cord and above the gut and supports the muscle blocks of the body in chordates.

nucleus A structure that is the controlling center of all cellular activity. It contains the CHROMOSOMES, the nucleolus, and the nucleoplasm.

O

ocellus A simple EYE—a cluster of photoreceptors—with a single lens system, found in arthropods and some other invertebrates.

ommatidium A single element of the COMPOUND EYE of insects and crustaceans.

operculum A cover of many animal structures, such as the lid of a NEMATOCYST; the horny cover with which some gastropods close their shell opening against desiccation or enemies; and, in bony fishes, the large bony plate that protects the GILLS.

ovary The egg-forming structure in a female or HERMAPHRODITE organism.

oviduct The tube that connects the OVARY to the UTERUS in female mammals and carries eggs out to the body in other vertebrates.

oviparous Describes animals that lay eggs that hatch after they are laid.

ovoviviparous Describes animals in which the EMBRYO develops in an egg membrane until it hatches inside the mother.

P

parasite An organism that lives on another, called a HOST, at its expense. Ectoparasites, such as fleas, live on the surface of their HOST; endoparasites, such as tapeworms, live internally.

parthenogenesis The development of an unfertilized ovum into a new individual. In some animals, this is the usual method of reproduction, but the pattern is often broken by a sexual generation which allows a recombination of GENETIC material within the animal's population.

pectoral girdle The BONES supporting the forelimbs in TETRAPODS, consisting of shoulder blades (scapulas), collar bones (CLAVICLES), and in some cases, CORACOIDS.

pelvic girdle The hip girdle of TETRAPODS composed, in most cases, of the ischium, the pubis, and the ilium. It supports the abdomen, the base of the tail, and the upper part of the leg.

pentadactyl Meaning "five fingered," this term refers to the limbs of early land vertebrates, which had five digits on each limb. This number has been reduced in many modern vertebrates, but the basic pattern is the same, so the term is still used.

peristalsis Waves of muscular contraction in tubular organs.

phenotype The appearance of a genetic trait in an organism.

pheromone A chemical, usually a scent, secreted by an organism which stimulates a particular response in another individual. Pheromones play a large part in the courtship and mating behavior of many animals.

photosynthesis The process by which green plants transform the light energy from the sun into chemical energy and use it to build carbohydrates. It occurs in the CHLOROPLASTS of plants.

phytoplankton The plant component of the PLANKTON. These unicellular organisms are the basic food for all ocean-dwelling animals and, through their PHOTOSYNTHETIC activity, they provide much of the free oxygen in the atmosphere.

placenta A temporary organ that develops in the UTERUS of most pregnant mammals through which the EMBRYOS are nourished, receive oxygen, and eliminate waste.

plankton The term covers all floating aquatic organisms. Many planktonic animals are LARVAL forms, but some are permanent members of the plankton. Most are microscopic, but a few, such as jellyfish, are large.

plantigrade The style of locomotion in TETRAPODS where the whole length of the foot is placed on the ground, as in man, most insectivores, and some reptiles.

pneumostome The RESPIRATORY pore of terrestrial mollusks, which is found behind the head, on the right side of the body.

poikilothermal Describes animals that are "cold-blooded." Their body temperature is not low, but matches that of their surroundings, fluctuating with the time of day or year.

polyembryony The formation of more than one EMBRYO from a single ZYGOTE by fission at an early stage.

polyp The fixed stage in the life of many coelenterates. In some, such as sea anemones and corals, it is the only adult form; but, in others, such as hydrozoa, it is the ASEXUAL reproductive phase.

proglottid A single reproductive segment of a tapeworm. It contains the male and female reproductive organs.

prognathous Projecting jaws, as found in baboons, apes, and some early humans.

protein A complex organic compound, formed from large numbers of AMINO ACIDS. Proteins play an important part in the formation, maintenance, and regeneration of tissues.

protoplasm The entire contents of a cell, which includes the CYTOPLASM and the NUCLEUS.

pseudopodium A temporary extension of a CELL caused by the directional flow of PROTOPLASM within it. It is the method of locomotion and feeding in ameboid cells.

pupa The stage between the LARVAL feeding form and the adult reproductive form in HOLOMETABOLOUS insects. During the pupal stage, the body tissues are almost completely broken down and reconstructed round a group of special CELLS.

R

rachis The main part of the shaft of a feather, which carries the vane made of barbs and barbules.

radial symmetry A circular body form in which there is no "head end" to the body and no concentration of BRAIN or nervous tissue. It is usually found among inactive animals.

radula The "tongue" of mollusks. It carries large numbers of horny teeth, which are used to tear food.

respiration The method by which an organism oxygenates its tissues and rids itself of unwanted carbon dioxide. In invertebrates, gases may diffuse through the whole of the outer surface or via TRACHEAE to the tissues. In larger animals, specialized organs have evolved, including GILLS for aquatic animals and LUNGS for terrestrial species.

retina The light-sensitive layer at the back of the EYE of vertebrates and cephalopods. It includes two types of cells—rods, which are sensitive to dim light, and cones, which are concerned with the perception of colors.

ribonucleic acid See RNA.

RNA (ribonucleic acid) The material in a CELL, composed of a single nucleotide chain, that organizes and governs PROTEIN synthesis.

royal jelly A highly nutritious secretion from the pharyngeal glands of young worker honey bees. It is fed to all the LARVAE in the colony for the first few days of their lives, after which most of them are switched to a pollen diet. Larvae which are to become queens continue to be fed solely on royal jelly.

rumination The method of digestion used by cud-chewing ungulates. They swallow their food unchewed, which goes into the rumen where it is partly broken down by BACTERIA. Later, it is returned to the mouth to be masticated completely and when swallowed, it by-passes the rumen.

S

scolex The head of a tapeworm, which is armed with hooks, by which it attaches itself to the wall of the HOST's intestine. The PROGLOTTIDS grow from the scolex.

sedentary Describes animals that remain in the same living area, such as nonmigratory birds which, even so, move through several acres of territory. Among the invertebrates, it describes animals such as oysters or barnacles which are fixed to one point of the substrate.

segmentation Repetition of a structural pattern along the length of an animal's body or appendage, seen most clearly in annelids and arthropods. Each repeated unit of the pattern is referred to as a segment.

sesamoid A BONE developed within a tendon, especially in mammals, such as the kneecap, and the "thumb" bone of the giant panda.

sessile Describes animals that are attached to the substrate, as are barnacles and oysters.

sexual dimorphism The differences between the male and female form of SPECIES. Often the differences are slight but sometimes they are so great that the male and female have been thought to be of unrelated species.

sexual reproduction The formation of a new individual from a fertilized ovum, or SYNKARYON. At conception, a male GAMETE fuses with a female gamete. Because during MEIOSIS there is considerable shuffling of GENETIC material, each gamete differs from all others. The individual formed as a result of sexual reproduction is genetically unique and even the offspring of the same parents differ from each other.

shoaling In fishes, the habit of forming large, tightly knit groups composed of individuals of the same age and size. The purpose may be safety—predators may think the group to be one large animal, too big to attack safely, and if an attack is made the large number of individuals may bewilder the predator.

skeleton The BONY internal support of vertebrates. It consists of the vertebral column, which supports the head at one end, a PECTORAL GIRDLE, to which the forelimbs are attached, and a PELVIC GIRDLE, which carries the hind limbs. Fishes have fins instead of girdles.

sonar See ECHOLOCATION.

species The basic unit of animal classification consisting of populations of organisms sufficiently similar to permit free interbreeding in the wild. (*See also* BINOMIAL CLASSIFICATION.)

spermatophore A packet of sperm produced by the males of some species in which there is internal fertilization but in general no intromittant organ. In many cases complex mating behavior patterns have evolved to ensure that the female is correctly positioned to pick up the spermatophore in her CLOACA.

spicules The skeletal support of sponges. They may be of a horny material as in bath sponges, silicaceous in glass sponges, or chalky as in calcareous sponges.

spinnerets Small raised organs at the hind end of a caterpillar or spider, through which silk is produced. Each one contains minute outlets through which the silk is forced in liquid form, in chains of molecules. The silk polymerizes on contact with the air.

spiracle In cartilaginous fishes, this small opening lies in front of the GILLS. In bony fishes, it is dorsally situated and is used to take water into the gill chambers. It is also a breathing pore found on the sides of insects which opens into the TRACHEAL system.

spore A dormant form that many protozoans adopt allowing them to withstand long periods of extremes of heat or cold which would be lethal at other times. They return to a normal, active form when conditions are appropriate.

stapes One of the three small BONES of the mammalian inner ear. Reptiles, amphibians, and fishes have a comparable bone, but are lacking the other two.

statocyst A more or less spherical organ of balance that contains a granule of hard material that stimulates sensory CELLS as the animal moves.

sternum The breastbone in land vertebrates, to which the ribs, CLAVICLES, and CORACOIDS (if any) are attached.

suspension feeder An animal which feeds on minute food particles strained from the water in which it lives.

swimbladder A gas-filled sac in the abdominal cavity of bony fishes which enables them to achieve neutral buoyancy. Its presence means that the animal need not spend energy in maintaining its level in the water, but rather can use all of its locomotive power in moving forward.

symbiosis An association of two or more different organisms which are mutually interdependent and cannot live alone.

synkaryon The diploid ZYGOTE NUCLEUS, formed by the fusion of GAMETIC nuclei.

synovial capsule A bag of connective tissue surrounding any free-moving joint such as a shoulder, elbow, or wrist. It contains synovial fluid, which lubricates the CARTILAGE coverings of the ends of the BONES.

T

tapetum A reflecting layer at the back of the EYE of some vertebrates, particularly those that are nocturnal. It enables the eye to make the fullest possible use of any available light.

tergum The thickened cuticle on the upper surface of a segment of an insect or crustacean.

testis The sperm-forming structure in a male or HERMAPHRODITE organism.

tetrapod Literally meaning "four feet," this term is used to denote any land vertebrate.

torsion The twisting of the visceral mass through 180°, which takes place in the LARVAL stage of gastropods, bringing the anus from its original posterior position into an anterior position above the head. The advantage appears to be that the mantle cavity then provides space into which the animal can retract its soft parts, within the protection of the shell, closed in many cases by the horny OPERCULUM.

trachea A breathing passage in insects and some arachnids. It conducts air from the SPIRACLES on the skin into the body, where it branches to form fine tubes called tracheoles, where gas exchange takes place. In higher vertebrates, the trachea is the windpipe.

trochophore A planktonic LARVA of polychaetes and some mollusks. Roughly spherical in shape, it has a ring of cilia round the body, the beating of which stabilizes the animal in the water and brings food to its mouth.

U

ultrasonic Describes high-frequency sound waves above the hearing of human ears, used by bats and whales in ECHOLOCATION.

unguligrade A method of mammalian locomotion in which the animal walks on the tips of its toes, which are generally protected by horny hoofs. Each stride is lengthened by the length of the hand or foot plus the toe bones, which are often greatly elongated. Many long-distance runners such as horses and antelopes are unguligrade.

uric acid A waste product formed from the breakdown of AMINO ACIDS. It is produced by animals such as insects, snails, reptiles, and birds, which need to conserve water rather than lose it in excreta.

urine Liquid waste produced by the kidneys (and stored in the bladder) which contains mainly the breakdown products of AMINO ACIDS.

uterus The muscular organ in female mammals and some other vertebrates in which the EMBRYOS develop.

V

vector A blood-feeding insect, such as a mosquito, which transmits parasitic organisms, for which it is the intermediate HOST, to another host in the PARASITE's life cycle.

veliger The LARVA of gastropods and some other mollusks, which develops from the TROCHOPHORE. The CILIATED region is drawn out into an elongated velum, which supports the animal like water skis. During the veliger state, TORSION takes place.

viviparous Describes the development of an EMBRYO nourished by the mother until it is born, when it is a smaller version of the adult.

Z

zooplankton The animal members of the PLANKTON. Many are LARVAE but some, such as jellyfish, include the largest members of the plankton.

zygote The fertilized ovum before it begins further development.

Index

Credits

The following have provided photographs for this book: Cover photo—Heather Angel; Michael Abbey/ Okapia 15; Kurt Amsler/Seaphot 62; Heather Angel 12, 19, 20, 25, 26, 37, 39, 42, 43, 45, 46, 52, 55, 59, 62, 66, 73, 74, 75, 80, 90, 93, 100, 102, 118, 119, 120, 144; Anthony Bavenstock/Seaphot 41; James Bell/Science Photo Library 9; S. C. Bisserot FRPS/Nature Photographers Ltd 107, 109, 117; Frank V. Blackburn/Nature Photographers Ltd 18; P. Boston/Natural Science Photos 113; Brinsley Burbidge/Nature Photographers Ltd 145; N. A. Callow/ Nature Photographers Ltd 61; Peter D. Capen/Seaphot 140; Dick Clarke/Seaphot 64; Colour Library International 45, 91, 92; Alain Compost/Bruce Coleman Ltd 97; Conor Craig/Seaphot 70; Richard Crane/Nature Photographers Ltd 139; Bill Curtsinger/Okapia 129; Tim Davis/Photo Researchers 10; Michael Dick/Animals Animals 141; Martin Dohrn/Science Photo Library 57, 74; Georgette Douwma/ Seaphot 69; Friskney Essex/Natural Science Photos 143; Douglas Faulkner/Okapia 131; Geoscience Features 35; Finnish Tourist Office 99; Ron and Christine Foord 53; David R. Frazier 9, 11; F. Gohier/Photo Researchers 31; Jon Green/Animals Animals 95; Gower Medical Publishing 32, 55, 72, 99; Dr Steve Gull/Science Photo Library 65; Eric Gravé/Science Photo Library 17, 20, 34, 36; Dan Guravich/Photo Researchers 129; James Hancock/Nature Photographers Ltd 130, 139; Margaret Hayman/Seaphot 59; A. Hayward/Natural Science Photos 76, 82; Carrol Henderson 118; Jan Hirsch/Science Photo Library 24; J. Hobday/Natural Science Photos 105, 106; David Hosking 25; James Hudnall/Seaphot 121; Alan Hutchinson Library 127; Johnny Johnson/Bruce Coleman Ltd 141; Edgar T.

Jones/Aquila Photographics 93; Masao Kawai/Orion Press/Bruce Coleman Ltd 112; G. C. Kelley/Photo Researchers 30; Nicholas Law 11, 133; Hugo van Lawick/Nature Photographers Ltd 123; J. Lawton Roberts/Aquila Photographics 92; Michael Leach/Nature Photographers Ltd 142; Ken Lucas/Seaphot 23, 33, 139; John and Gillian Lythgoe/Seaphot 126; R. L. Matthews/Seaphot 141; C. Mattison/Natural Science Photos 79; L. C. Marigo/Bruce Coleman Ltd 110, Nature Photographers Ltd 49; Peter Newark's Western America 134; Greg Ochocki/Photo Researchers 31; Okapia 60, 101; W. S. Patou/Nature Photographers Ltd 139; J. M. Pearson/Biophotos 19, 132; Howard Platt/Seaphot 128; K. R. Porter/Science Photo Library 15; Jurge Provenza/Seaphot 110; W. J. von Puttkamer/Alan Hutchinson Library 104; Ian Redmond/ Seaphot 61, 114; Hans Reinhard/Bruce Coleman Ltd 94; Brian Rogers/Biofotos 91; Rod Salm/Seaphot 78; Scala III; Peter Scoones/Seaphot 70,71; Jonathan Scott/Seaphot 22, 27, 28, 29; Seaphot 46, 48, 49, 53; David Sewell/Nature Photographers Ltd 139; Silvestris/Meyers 124; Silvestris/Wothe 125; Dr Nigel Smith/Alan Hutchinson Library 104, 135; Sinclair Stammers/Science Photo Library 48; Tony Stone Images 6, 8, 10, 36, 63, 68, 73, 81, 83, 93, 95, 96, 100, 109, 123, 125, 133, 135, 137, 143; Soames Summerhays/Biofotos 12, 128, 139, 145; Herwath Voigtmann/Seaphot 67; John Walsh/Science Photo Library 13, 40,·51; P. H. and S. L. Ward/Natural Science Photos 79, 119, 131; Gary Weber/Aquila Photographics 77; Curtis Williams/Natural Science Photos 87; Robin Williams/ Gower Medical Publishing 99; York Museum 139.